The Practical Pendulum Book

'This is a fascinating, unpretentious, splendid little book. It fills a gap on everyone's shelf of psychic books.'

— *Spiritualist Gazette*

THE PRACTICAL PENDULUM BOOK

with instructions for use

and

thirty-eight pendulum charts

D. JURRIAANSE

Aquarian/Thorsons
An Imprint of HarperCollins*Publishers*

Publisher's Note: ***The information presented here has been obtained through the author's research conducted over many years. Other research may indicate different results or different opinions. The author does not intend this book to be used to treat physical conditions that should be handled by a competent professional.***

The Aquarian Press
An Imprint of HarperCollins*Publishers*
77–85 Fulham Palace Road,
Hammersmith, London W6 8JB

First published in Dutch by Uitgeverij Schors,
Amsterdam, Holland, as *Praktisch Pendelboek*, 1984
First published in English by Samuel Weiser Inc.,
York Beach, ME 03910, USA, 1986
This edition published by The Aquarian Press, 1987
10 9 8 7 6 5 4

A catalogue record for this book
is available from the British Library

ISBN 0 85030 609 4

Printed in Great Britain by
Woolnough Bookbinding Limited,
Irthlingborough, Northamptonshire

Pendulum Basics

A pendulum can be made of gold, silver, copper, bronze, wood or glass, but when you are just beginning it is best to use a pendulum made of the material related to your sign of the zodiac. The best shape to buy is the type that tapers to a point; some examples are shown in figure 1 on page 2. It is advisable not to use a very heavy pendulum for your first experiments! It should weigh less than half an ounce, and the chain or silk thread on which it hangs should be 10 to 12 inches long. In fact, it should be roughly the same length as your forearm (measured from the elbow to the fist).

Beginners should try the following position. Place the elbow firmly on a table and hold the pendulum an inch or two above the object on which it is to be used. You must avoid any unnecessary tension in the muscles, particularly those of the arm and hand. Then try to concentrate without thinking of anything specific. The best way to do this is to imagine you are waiting for a particular event without having any idea what this event might be. Your body and mind should be relaxed and passive and any superfluous energy should be avoided.

Use the pendulum with the right hand, even if you are left-handed. Rest your left hand lightly on the back of your right hand, spreading the thumb and fingers slightly and making sure that they do not touch each other—the fingers' own polarity can create disturbances.

Now that you are completely prepared to start using the pendulum, place an object made of gold beneath it and wait patiently to see what will happen. The first movements will be a strong experience, for you will be aware and feel clearly that it is not you, or your arm or finger that causes these movements, but that there are forces acting together. Become aware of the fact that you are part of an infinite cosmic force field and you will

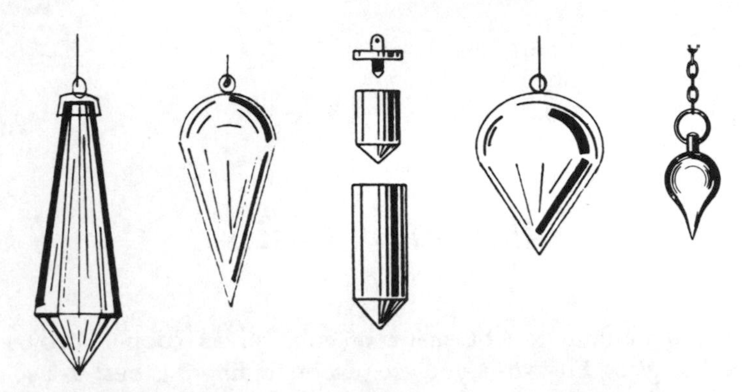

Figure 1. The pendulum. The best shape to look for is one that tapers to a point.

realize that you can actually feel this force field. As you become more experienced, you will not have to keep so strictly to these rules.

The pendulum can move in three different ways: with a circular movement, elliptical movement or in a straight line. There is a marked difference between these movements. (See figure 2.)

When you are using the pendulum, keep your eyes on the point. Make sure that your imagination does not lead your eyes to follow an imaginary pattern, since this could influence the results. Any object on which you wish to use the pendulum should also first be purified from extraneous radiation that might also play a role. This can be done by placing the object in running water for a while. However, the same effect can be achieved by magnetizing the object with the left hand. To do this, hold the left hand a little bit above the object and take a few deep breaths. Every time you breathe in, imagine that all external influences that could have become attached to the object are sucked up while you breathe in through the magnetic pole of the left hand. When you have done this a few times, shake the left hand as though you are shaking off a few drops of water. If the pendulum now *hangs quite still* over the object, you will know that all the extraneous influences have been removed.

Obviously, you will understand that in view of these extraneous influences you must take care not to let your pendulum fall into strange hands. For this reason, carry it with you wherever you go, preferably wrapped in a silk cloth.

Before you really start to use the pendulum, you must first charge it. This is done by holding it in the right hand and breathing on it a few times. While doing this, imagine as intensely as possible that you are filling the pendulum with your own life force. *Do not forget that your imagination is the most powerful tool at your disposal.*

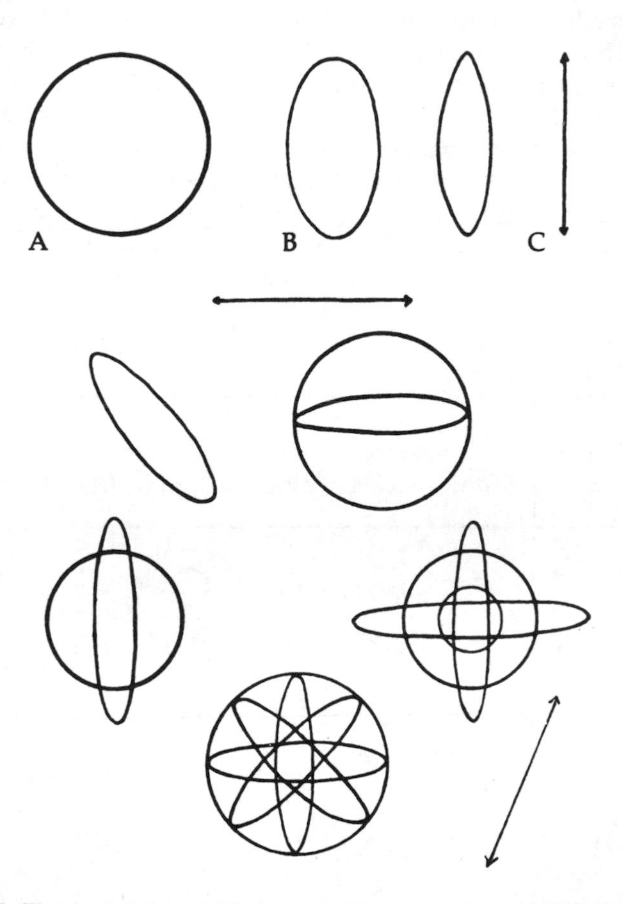

Figure 2. The pendulum will move in three basic ways: a) circular movements; b) elliptical movements; and c) in a straight line. There are many variations on these basic movements, as shown in this illustration.

Experimenting with the Pendulum

Our first series of experiments is concerned with using the pendulum on minerals. This is done mainly to determine the type of figure described by the pendulum—circle, ellipse or straight figure—as shown in Table 1. However, it is also done to determine which degrees it describes between 0° and 360°. (See the pendulum chart on dynamic values.)

Now take the pendulum in your hands. We will assume that it is made of copper or gold. Place an object made of gold beneath it. You will notice that the pendulum will slowly start to move in *a wide circle to the right.* The width of this circle depends on the purity of the gold. The same figure is obtained for radium, uranium, calcium, copper, manganese and platinum as for gold. However, if you use the pendulum over a silver object you will find that the pendulum *turns to the left and tends to describe an elliptical figure.* The same figure is also made with iron, nickel, aluminum, sulphur, lead, cadmium, and a number of other minerals.

Table 1. Using the Pendulum on Minerals

Pendulum Material	Used over: Iron	Nickel	Zinc	Copper	Lead	Alloy
Copper	⤢	↻	↔	↻	↻	↻
Magnesium	⤢	⤢	↕	⬭	↺	↕
Stone	⤢	↕	⬯	↔	↻	↻
Wood	↕	⤢	⤢	↔	↻	↻
Crystal	↻	⤢	⤢	↻	↻	↻
Steel	↕	⤢	⬯	↔	○	○
Silver	⤢	⤢	⤢	↔	○	○

I have deliberately started by using the pendulum with gold and silver, as these will reveal the movements corresponding to the most important polarities: gold-silver; sun-moon; male-female; mind-matter; positive-negative; dynamic-receptive. *A clockwise rotation indicates a male radiation; a counterclockwise rotation indicates a female radiation.*

When you extend these experiments to inorganic substances, chemical compounds, fossils, plants, perfumes, herbs, foodstuffs and so on, there is an almost infinite series of possibilities. However, you should only try these things when you feel that you are in control of your pendulum and you know what it is telling you.

The next step is to use the pendulum with colours. Paint a few sheets of paper with poster paints or use coloured paper and determine the patterns described by the pendulum over the main colours. Compare your results with those in Table 2. Make notes about your own results and compare them with your findings the next time you do the same tests.

When you start to feel confident enough about the results of your experiments you can go one step further. You no longer

Table 2. Using the Pendulum on Colours

Colour	Pendulum Material:					
	Copper	Magnesium	Stone	Wood	Crystal	Steel	Silver
Bright blue	↔	↺	↻	↕	⤢	↔	○
Dark blue	↻	↻	↔	↻	↕	↺	⤢
Green	↺	↺	⤡	⤡	↻	↺	↻
Red	↻	⤢	↻	↻	↺	↻	↻
Black	↺	↺	↔	↕	↔	↻	↺
Carmine	⤢	↻	↕	↻	↻	↺	↔

have to rest your arm on a table when you are using the pendulum, but can move freely. Now you can develop from so-called "figure pendulum exercises" to the so-called "degree pendulum exercises." The chart of dynamic values is accompanied by a classification of degrees and the four points of the compass (see Chart 5). Every object has its own degree, which you can get to know through your pendulum. The Table of Dynamic Values, facing Chart 5, shows the main materials and substances so that you can orient on these. However, they only serve as a guideline.

We will begin by using the pendulum with people. It is possible to use the pendulum to examine the different parts of the body in which the strength of radiation gives an indication of sex, character, and the physical and emotional condition of the individual. In this way you can discover whether the polarity might be incorrect or whether the psyche is not in equilibrium. For example, when more elliptical than circular patterns are described for a man, you can assume that his disposition has a more feminine than masculine character. The pendulum will react most strongly to the sexual organs; the movement of the pendulum will show almost immediately whether there is a sun or a moon force. To find the correct polarity you must use the pendulum over the person's thumb. Normally the pendulum will describe wide circles. For the left thumb there will be clockwise circles; for the right thumb they are counterclockwise.

• • •

The pendulum is also extremely useful for obtaining reliable information regarding a person's spiritual development. If you hold the pendulum over the bridge of someone's nose, you can discover some of the main drives influencing his behaviour. For this you should concentrate on the head and nervous system. Normally the pendulum will move in a clockwise direction, as it does for the heart. You can draw various conclusions from the nature of the circles. An essential factor in this is *the number of circles* described by the pendulum, as well as their *size and nature.* In some cases the pendulum will describe *a triangular pattern over the head and heart.* In this triangle the longest point is upwards. This indicates a *highly developed spiritual life.* (See Table 3.)

Also try holding the pendulum over a person's chest; you will feel the force emanating from the lungs. *It is often possible to*

Table 3. Using the Pendulum over the Body

cortex, face	○ (△)	intestines	⬭
cerebellum, lower jaw, neck	○	abdominal organs	↔
top of right lung	⤡	right leg, thigh	↕
bronchi	↕	right leg, knee	○
top of left lung	⤢	right leg, lower leg	↓
heart	○ (△)	left leg, thigh	↓
lungs	○	left leg, knee	○
stomach	◯	left leg, lower leg	↓
liver	⬯	feet	○
solar plexus	◯	life force	small or large circle

distinguish diseased areas from the fact that the pendulum stops moving entirely. When the pendulum reacts with *pulling movements* toward a point diagonally above and to the right of an organ, it can, in some cases, indicate cell destruction of the tissue of this organ.

The heart can also be examined in the same way. In healthy people there is an *even, circular, clockwise movement.* Whenever this is broken by restless movements, tremors or pulling of the pendulum, this indicates either a nervous condition and tension or damage, or a complaint in the organ itself.

You can also try the pendulum over *the stomach* area and the *solar plexus.* The usual *large, broad circles* clearly reveal the importance of this area in the nervous system and provide some insight into the function and effect of the nervous system. Keep the pendulum above the *body* and you will have *horizontal, wide elliptical movements.* A little lower down above *the lower abdomen,* the pendulum will move in *straight lines across the body,* indicating the pure, vegetative aspect of this area. Although the pendulum is supposed to react with straight lines in this area, it should not react this way for any other part of the body. Straight lines elsewhere would be indicative of something wrong. If the line is *not straight across the body* but moves diagonally, this indicates a *sexual deviation.*

You can also use the pendulum over photographs, handwriting or even the shadow of another person. You will always obtain the results outlined above.

It is also possible to find out a great deal about a person's character. There are many different opinions regarding the examination of a person's character using a pendulum. This is not very surprising when you consider how many different vibrations and degrees of vibration constitute a personality.

Hold the pendulum above the chest and concentrate on letting it tell you about a person's character. At first there will be the usual elliptical and circular patterns. If these are small and narrow, the person concerned has an inhibited character. If there is an absolutely straight line, selfishness and an obstinate character is indicated. If the line is long and the pendulum moves with some force, the person concerned is energetic and purposeful, but will not take other people into account and may even be rather ruthless. If the pendulum moves in quite wide circles and ellipses, this indicates a strong, constructive character.

Sometimes the figures start off narrow and only gradually become wider. This means that the person concerned has broken through the narrow limitations of his existence and has embarked on a period of spiritual growth. If the circles start off large and gradually become smaller and tighter, this indicates a development in the opposite direction. If the figures described by the pendulum become *horizontal ellipses,* this is indicative of tolerance, a sympathetic character and an understanding of human weakness. If the lines *go towards the feet,* this shows an interest in the good things of the earth, and in many cases it also indicates a materialistic streak. If the circle is slightly flattened at the top so that the cerebral cortex is not included, this reveals a lack of intelligence and the absence of spiritual interest. In this case the pendulum will only describe one sort of figure: for men, only neutral circles, and for women, only neutral ellipses. Both these figures point to purely vegetative functions.

One thing should never be forgotten: In all these experiments, thought vibrations produced either by yourself or by others can be extremely disruptive. It is obvious that the pendulum is extremely sensitive to thought vibrations—therefore it is important to work carefully to avoid the possibility of influencing the pendulum.

At this point another important factor should be considered: All the moods, feelings and vibrations of other people will be passed to you. You will be able to feel their influence clearly in the arm holding the pendulum. Illnesses can also be transmitted through the pendulum. You must therefore never forget to protect yourself in the correct way with rhythmic breathing. For example: When you use the pendulum with people who are in harmony, it will describe large, wide, regular circles, and often the thread or chain is not even long enough to describe the whole circumference. But when using the pendulum with people who are not in harmony you can often feel a sense of revulsion and rejection; your arm might start to pull away, or you might feel tired, or the pendulum might fall out of your hand. In these cases it is important to start breathing rhythmically in order to get rid of the negative vibrations that have taken hold of you. A strong downwards pull on the pendulum when you use it over a person is an unfavourable sign. This negative impression is even stronger if the pendulum is describing a line that runs east-west. Unfavourable tendencies in a person's character are also indicated when the pendulum changes its direction of rotation one or more times.

Continue to practice with experiments of your own. When you are using the pendulum to find out about character it is particularly important to come to conclusions that are valid for you personally and put your mind at rest. An interesting experiment might be to try analysing your own character! Write down the characteristics you discover and see which figures the pendulum uses to indicate them. By practicing in this way you can achieve wonderful results and learn to know yourself.

It is clear that the pendulum can also be used to determine sympathy or antipathy—attraction and repulsion—between people, animals, plants and minerals. *A circle that embraces* the people or objects indicates sympathy, closeness and cohesion; *a line that divides* the people or objects denotes rejection and antipathy.

If you want to find out whether two people suit each other you can hold the pendulum between their two right hands. The result will provide you with the answer to your question. In the same way, you can also find out what things are good for you and which are not. Put your left hand on the table and place some tobacco next to it. Now hold the pendulum between them and concentrate on the attraction and repulsion between them. If you can achieve *positive results* with the substance concerned the pendulum will describe *large circles,* sometimes encompassing the whole hand. In the opposite case, the pendulum will move in a *straight dividing* line between the hand and the object concerned and it is advisable not to use any of that substance. It is possible to use the pendulum in the same way to distinguish the type and usefulness of various medicines.

Using the pendulum to determine illness is very specialized and to do it well it is necessary to practice very intensely. It is only possible to make a few short remarks on this subject here. For example, hysteria will cause the pendulum to move in nervous jumpy lines. Weak nerves are indicated by east-west lines that decrease in size and eventually come to a standstill.

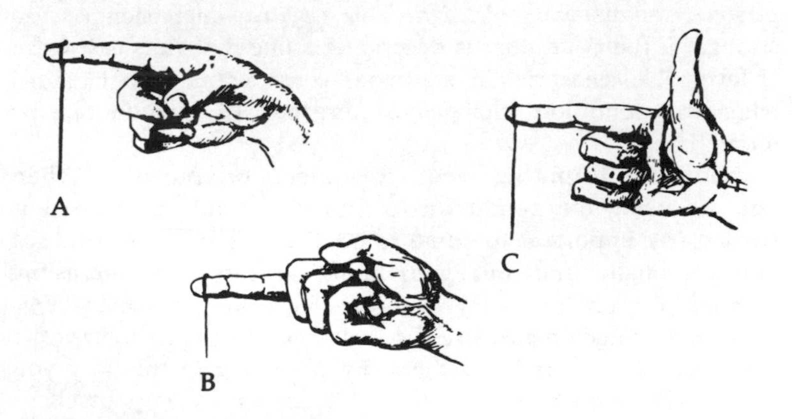

Figure 3. Holding the pendulum. Positions A and B are best when using a conical or cylindrical pendulum.

When the pendulum describes nervous lines over the forehead and seems to hit against a solid wall at the ends of the lines, a limited intelligence is indicated.

We can also examine the use of the pendulum with different gems. Gems contain various magic forces that can influence people in positive or negative ways. Obviously it is important to know which gem goes with a particular individual.

In this day and age, synthetic gems are made that are barely distinguishable from the real thing. Their weight, hardness and overall characteristics can even fool a professional jeweller! The pendulum can provide a solution. Concentrate on the life force of the gem. This force will be clearly indicated for real gems, while the pendulum will remain motionless over fake gems.

In principle you should use the procedures in all the experiments described above. It is only by using the pendulum as proscribed that you will become really experienced. After you've gained some experience you can choose a more unusually shaped pendulum, and change the position of your fingers to discover how to reach the highest degree of sensitivity so that you can feel the slightest vibrations. Figure 3 shows a number of possibilities for holding the pendulum. To feel the bigger vibrations, it is best to use the finger positions A and B in figure 3 with a conical or cylindrical pendulum. The pendulum should be fairly large and heavy in order to sense the force fields concerned. The more subtle the vibrations, the finer the pendulum should be and the more finely the finger position should be tuned into cosmic rays.

Pendulum Charts

Gemstone Significance

1. Lapis lazuli: friend stone
2. Chrysoprase: for a sore throat
3. Olivine: against evil influences
4. Jasper: strengthens the stomach, liver and kidneys
5. Garnet: strengthens the will
6. Onyx: strengthens the sense of hearing
7. Malachite: strengthens the heart, eyes and nerves
8. Rutile quartz: for bronchitis and asthma
9. Rock crystal: strengthens the back and solar plexus
10. Opal: stone of ill luck, attracts jinxes
11. Tiger eye: against fear
12. Diamond: inspires self-confidence
13. Sapphire: inspires purity, serenity and loyalty
14. Beryl: strengthens the sense of sight and the liver
15. Emerald: strengthens the intelligence and the memory
16. Aquamarine: prolongs youth, enhances marital bliss
17. Hyacinth: stimulates dreaming, anti-allergic
18. Ruby: opposes decay, fever and melancholia
19. Coral: prevents anemia
20. Turquoise: prevents accidents, fulfills wishes
21. Topaz: prevents enmity and false friends
22. Amethyst: for drunkenness, sleeplessness and migraine
23. Amber: stimulates the growth of teeth
24. Moonstone: promotes fertility, dependency and provides inspiration and love

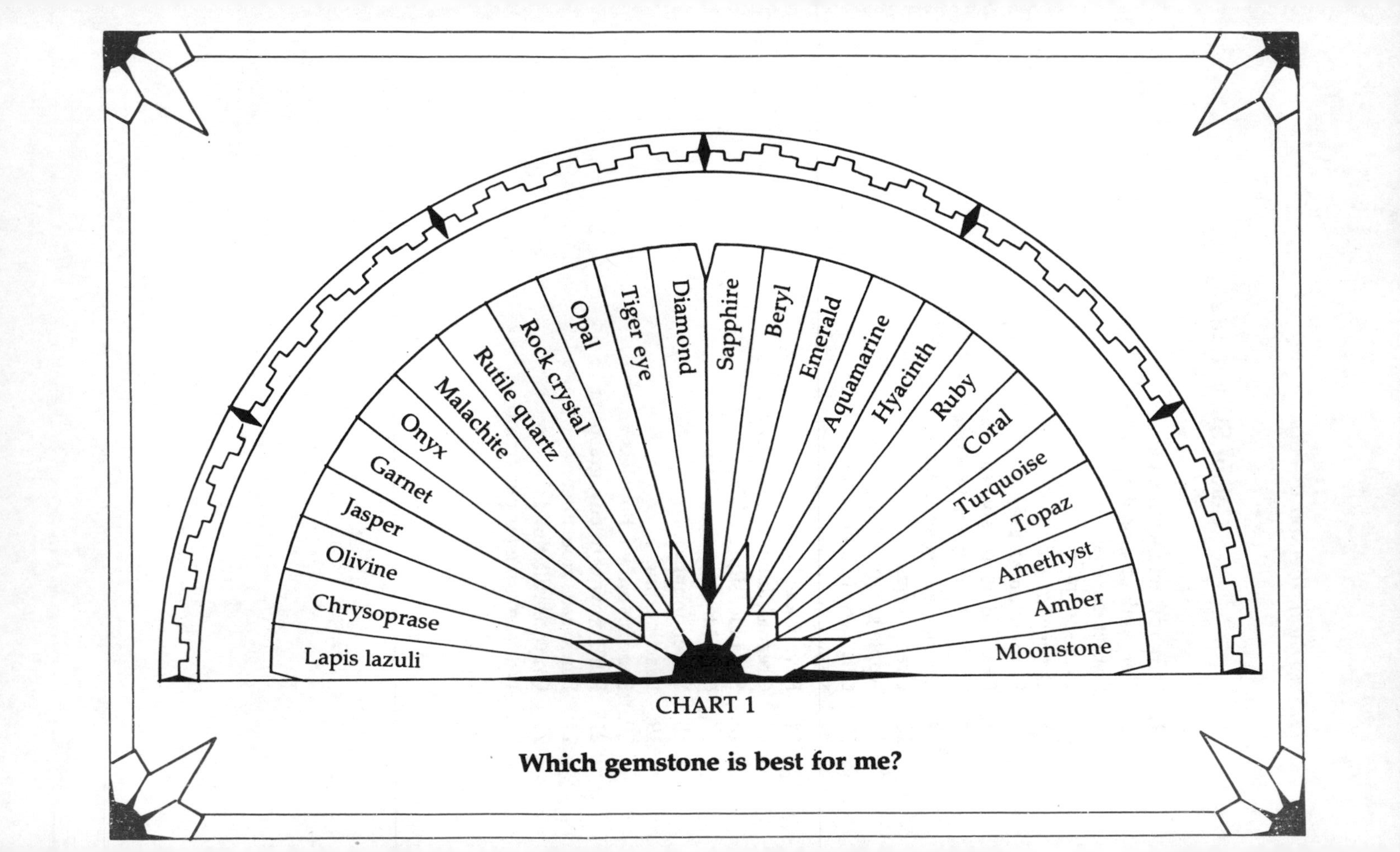

CHART 1

Which gemstone is best for me?

Colour Significance

1. Dark red: temperamental
2. Fiery red: passionate
3. Scarlet: sensual
4. Crimson: merry
5. Orange: proud
6. Salmon: immature
7. Ochre: devious, sly, jealous
8. Pale yellow: mentally active
9. Lemon yellow: spiritual
10. Green: lovable
11. Apple green: musical
12. Olive green: melancholy
13. Navy blue: decent, respectable
14. Royal blue: loyal
15. Greenish blue: self-confident
16. Sky blue: superficial
17. Gray blue: superstitious
18. Violet: intuitive
19. Lilac: healing
20. Reddish brown: hopeless
21. Dark brown: greedy, grasping
22. Black: sad, mourning
23. Gray: fearful, timid
24. White: a drive for perfection

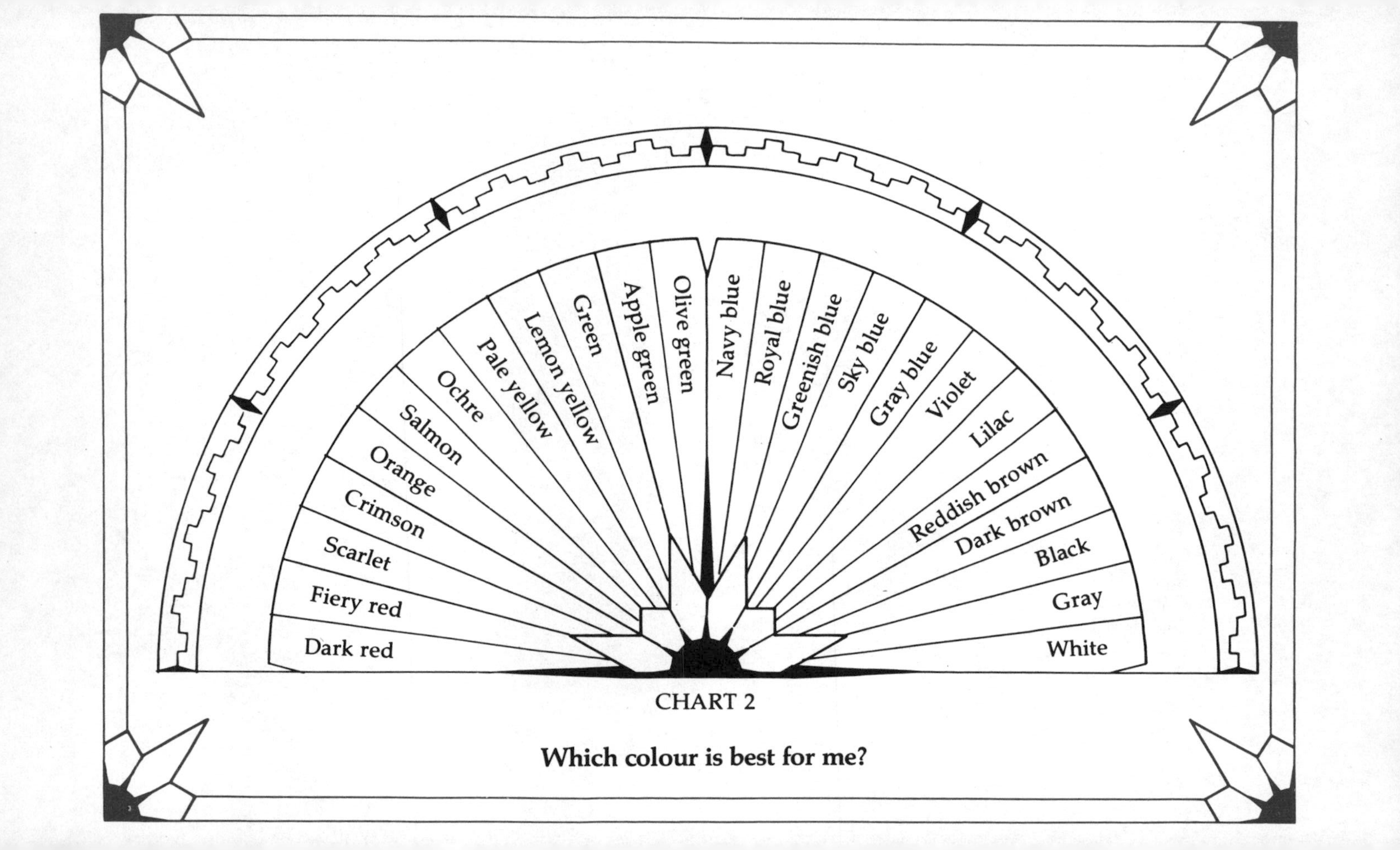

CHART 2

Which colour is best for me?

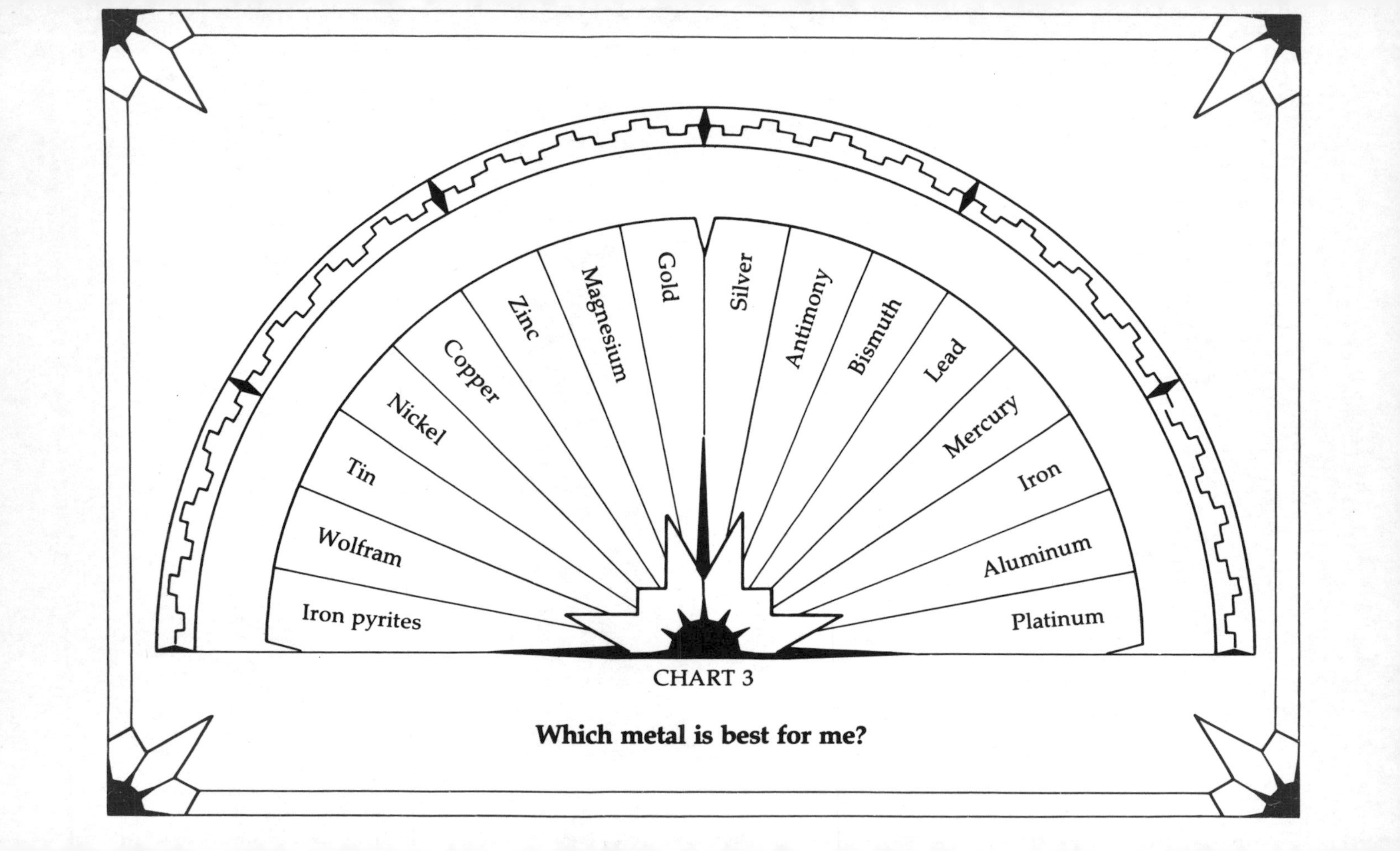
Iron pyrites
Wolfram
Tin
Nickel
Copper
Zinc
Magnesium
Gold
Silver
Antimony
Bismuth
Lead
Mercury
Iron
Aluminum
Platinum
CHART 3
Which metal is best for me?

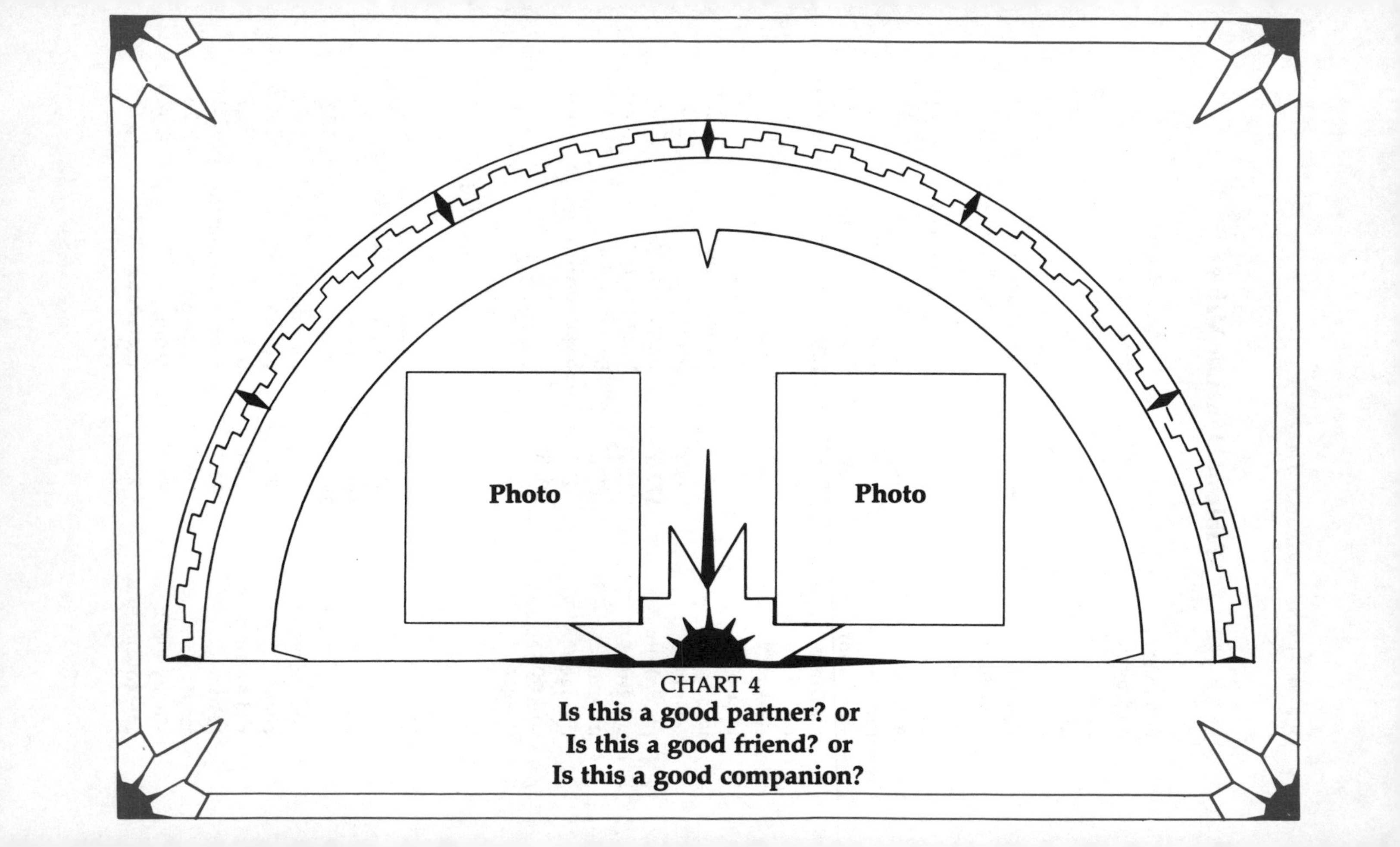

CHART 4
Is this a good partner? or
Is this a good friend? or
Is this a good companion?

Table of Dynamic Values

Elements:

hydrogen	0°
gold	0°
diamond	5°
silver	45°
zinc	67.5°
palladium	90°
uranium	100°
copper	112.5°
iridium	120°
tin	125°
aluminum	130°
platinum	135°
nickel	140°
cobalt	145°
lead	150°
iron	157.5°
selenium	175°
phosphorus	177.5°
sulphur	180°
mercury	270°
nitrogen	270°
chlorine	355°
oxygen	360°

Precious stones:

diamond	5°
chrysolite	247.5°
opal	247.5°
pure rock crystal	267.5°
common quartz	270°
chalcedony	270°
red tourmaline	280°
garnet	292.5°
jasper	295°
smoky topaz	297.5°
fine tourmaline	290°
lapis lazuli	290°
sapphire	322.5°
beryl	290°
emerald	297.5°
aquamarine	300°
emerald, bright green	302.5°
chrysoberyl	305°
hyacinth, pale red	305°
hyacinth, dark red	315°
gray tourmaline	272.5°
black tourmaline	290°
ruby	312°, 315°

Colors:

cobalt blue	315°
indigo blue	255°
bright blue	217.5°
bright orange	222.5°
bright red	215°
bright violet	230°
bright brown	240°

Miscellaneous:

ice, snow	202.5°
rainwater, distilled water	180°
brass	ca. 105°
half silver, half copper	77.5°
cast iron	150°
steel	152.5°
lodestone	330°
humus	292.5°
urine	270°
human excrement	315°
caffeine	265°
tannin	267.5°
morphine	292.5°

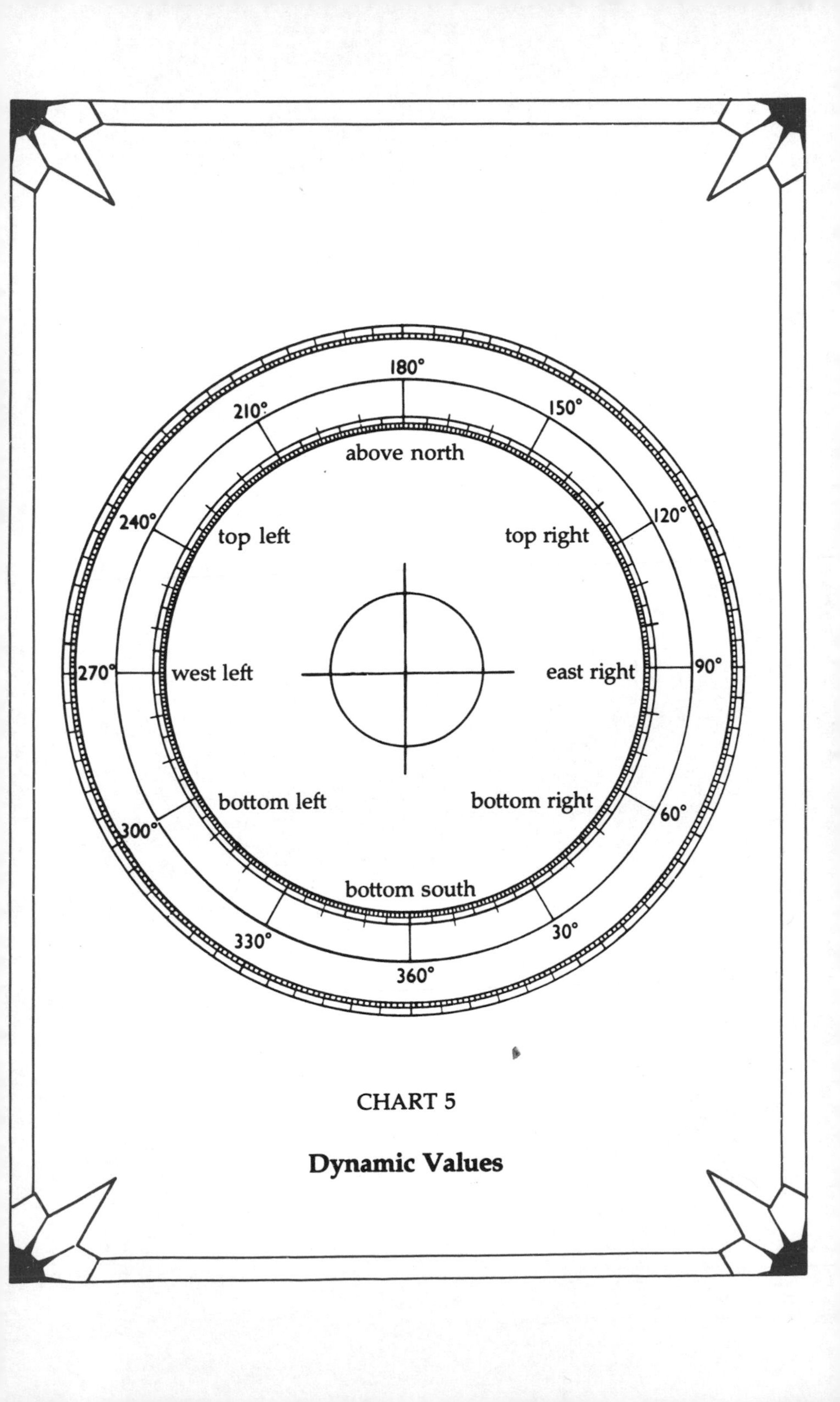

CHART 5

Dynamic Values

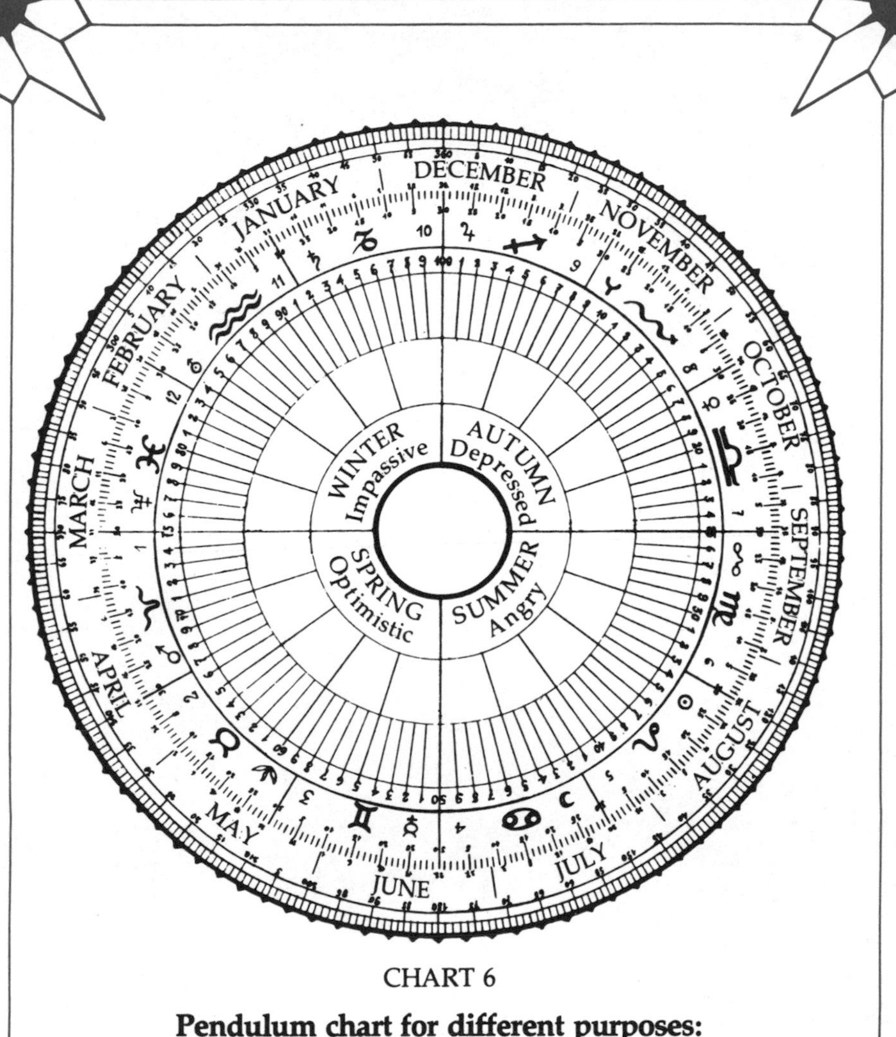

CHART 6

Pendulum chart for different purposes:

1. Diagram of months with data
2. Diagram of seasons
3. Diagram of temperaments
4. Astrological diagram
5. 100 and 360 degree diagram rotating in a clockwise direction

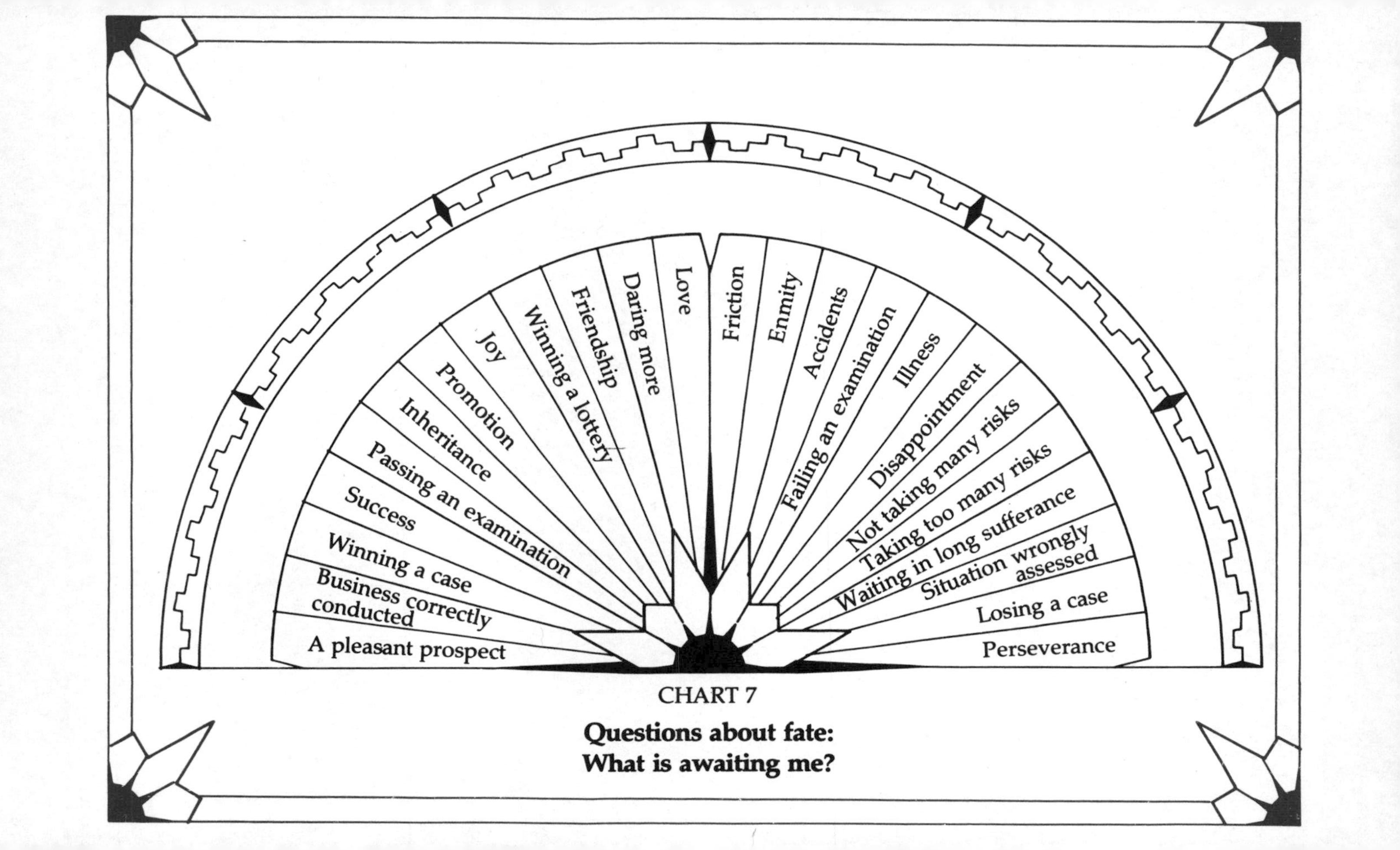

CHART 7

Questions about fate: What is awaiting me?

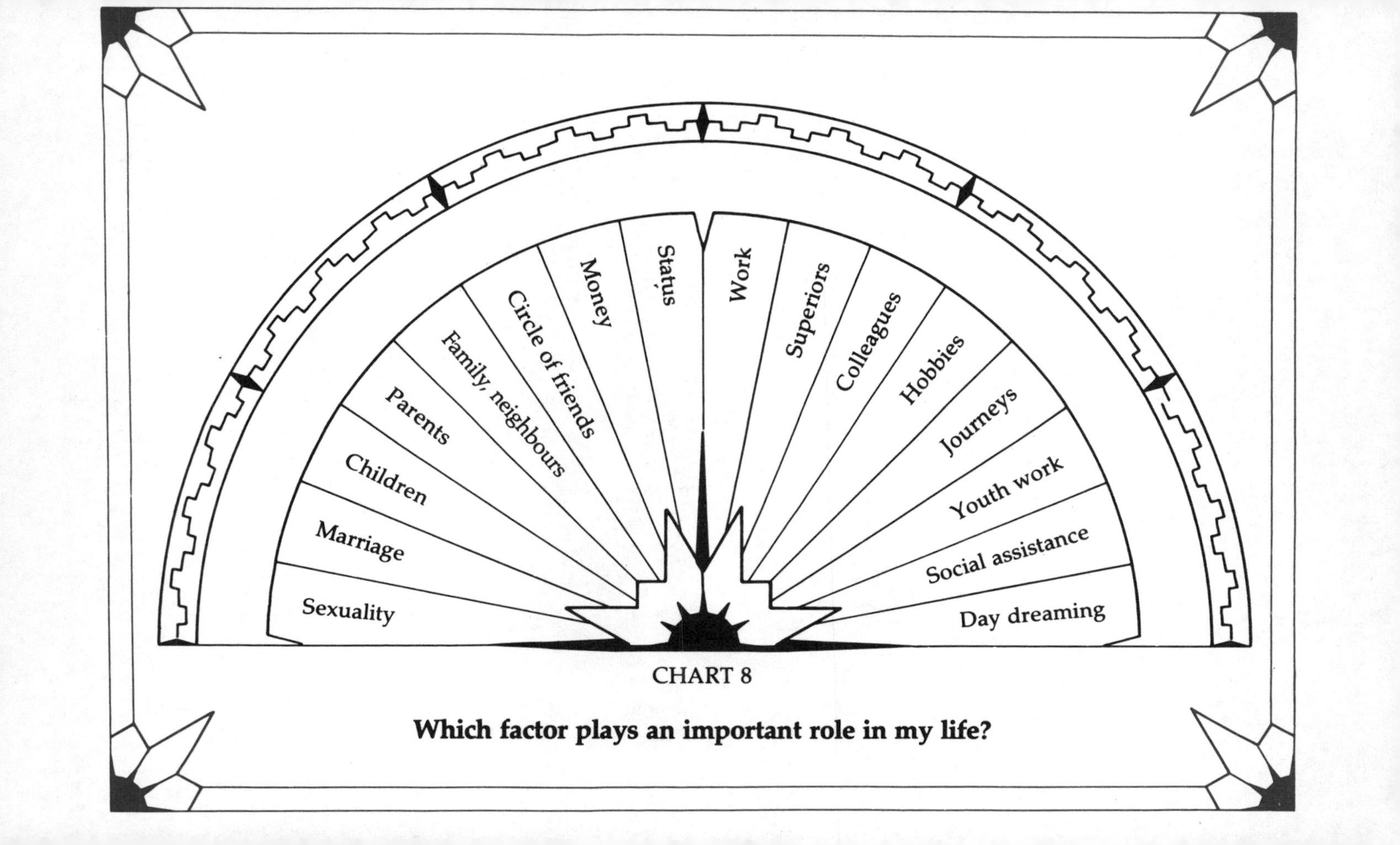

CHART 8

Which factor plays an important role in my life?

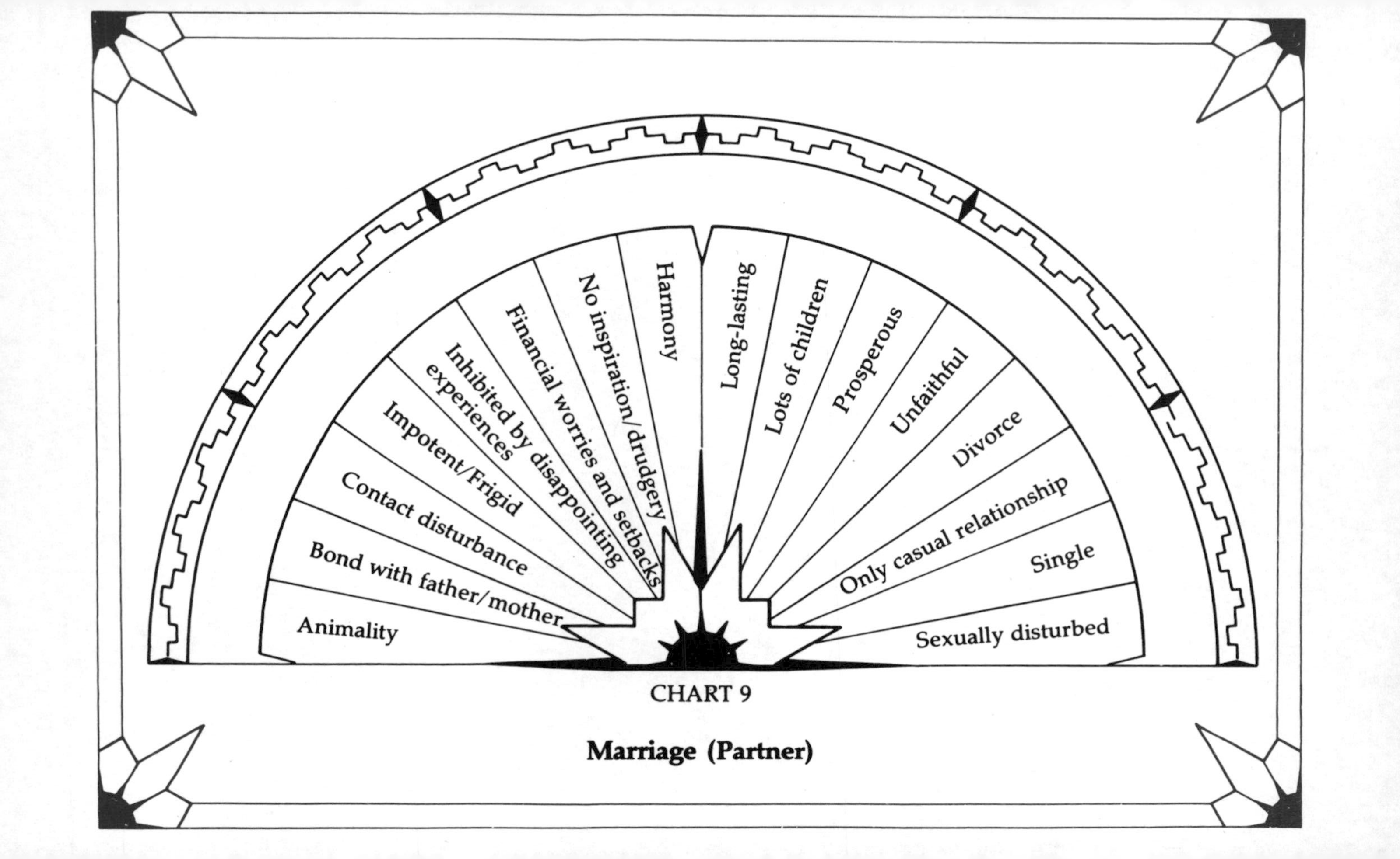

CHART 9

Marriage (Partner)

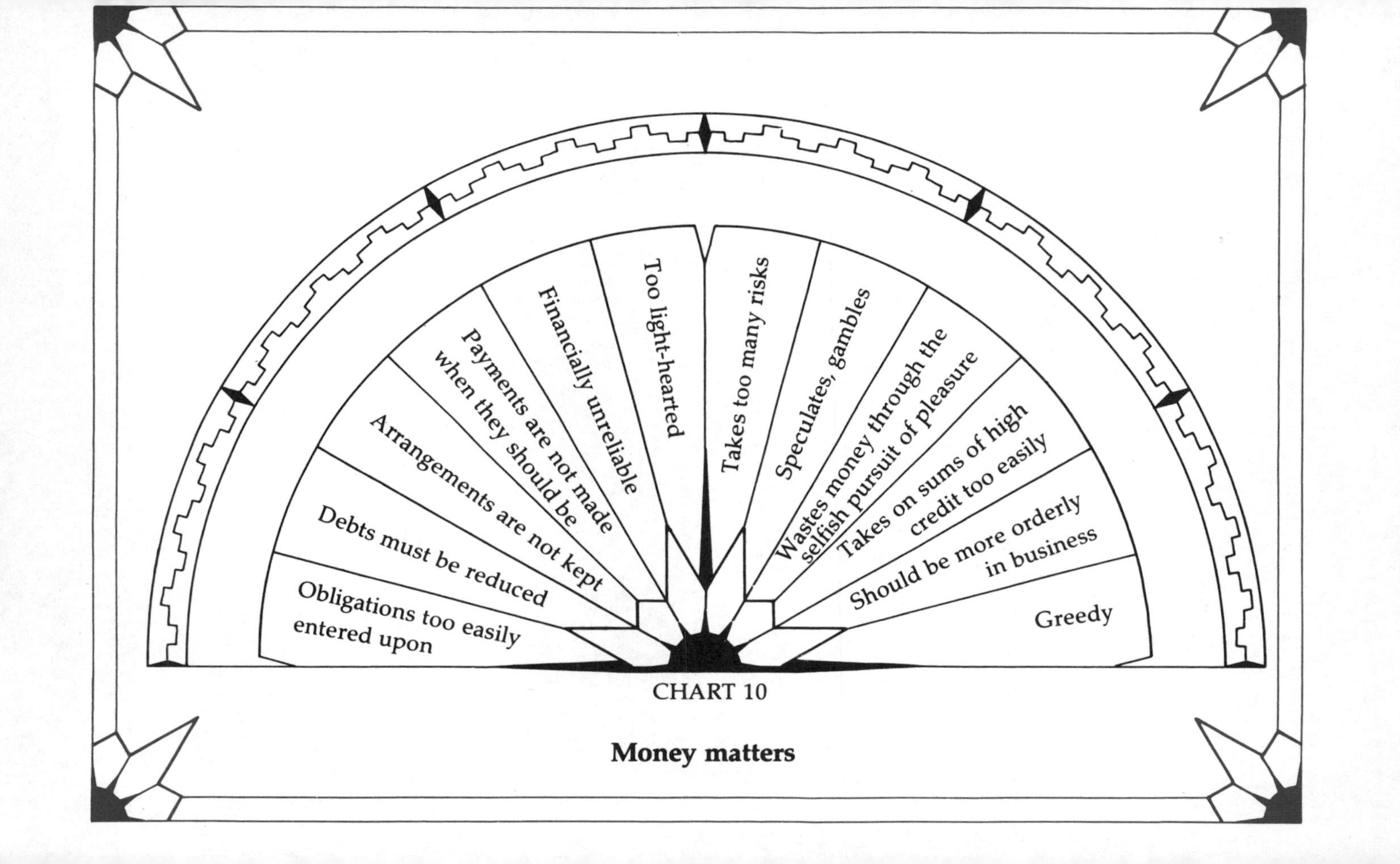

CHART 10

Money matters

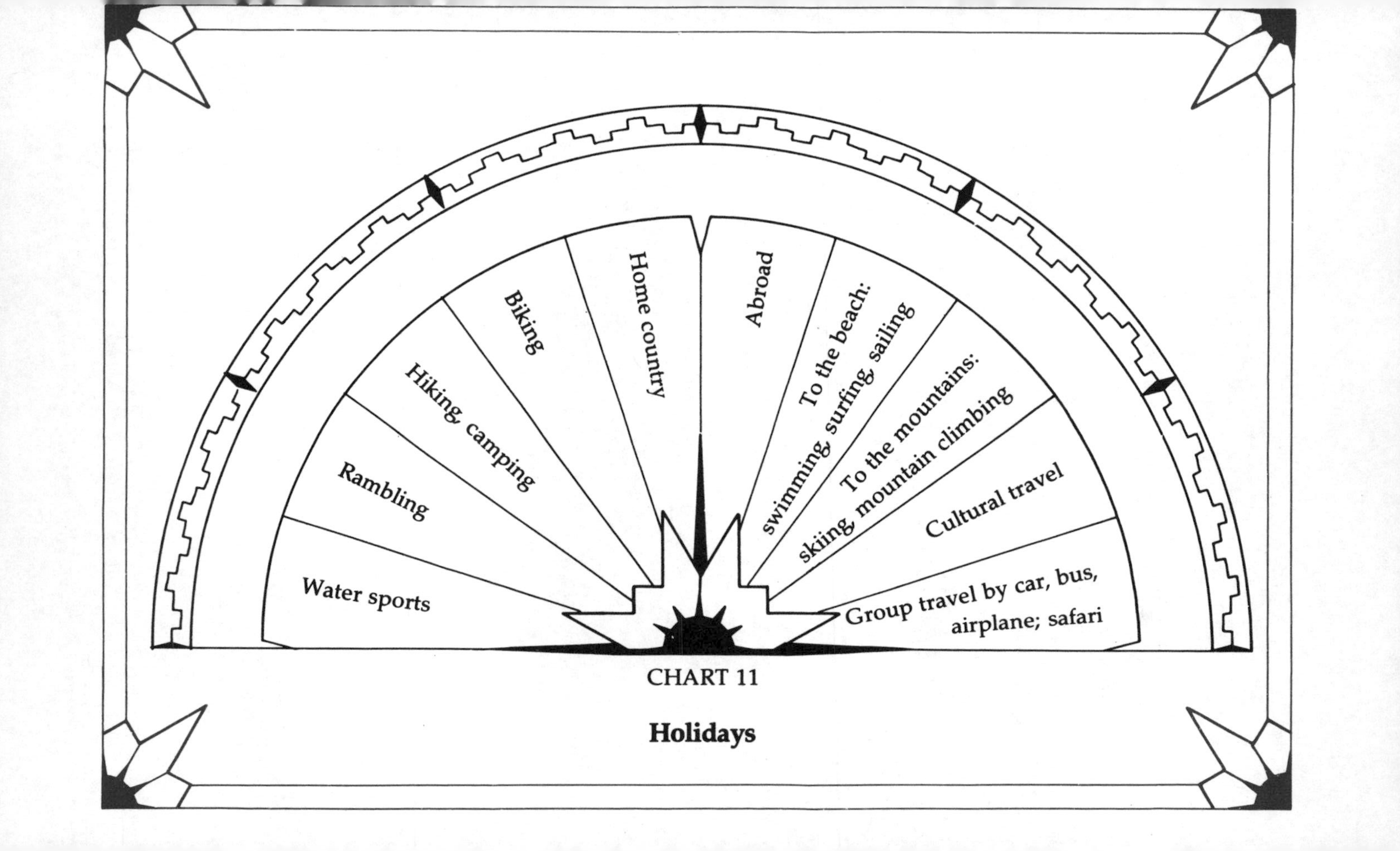

CHART 11

Holidays

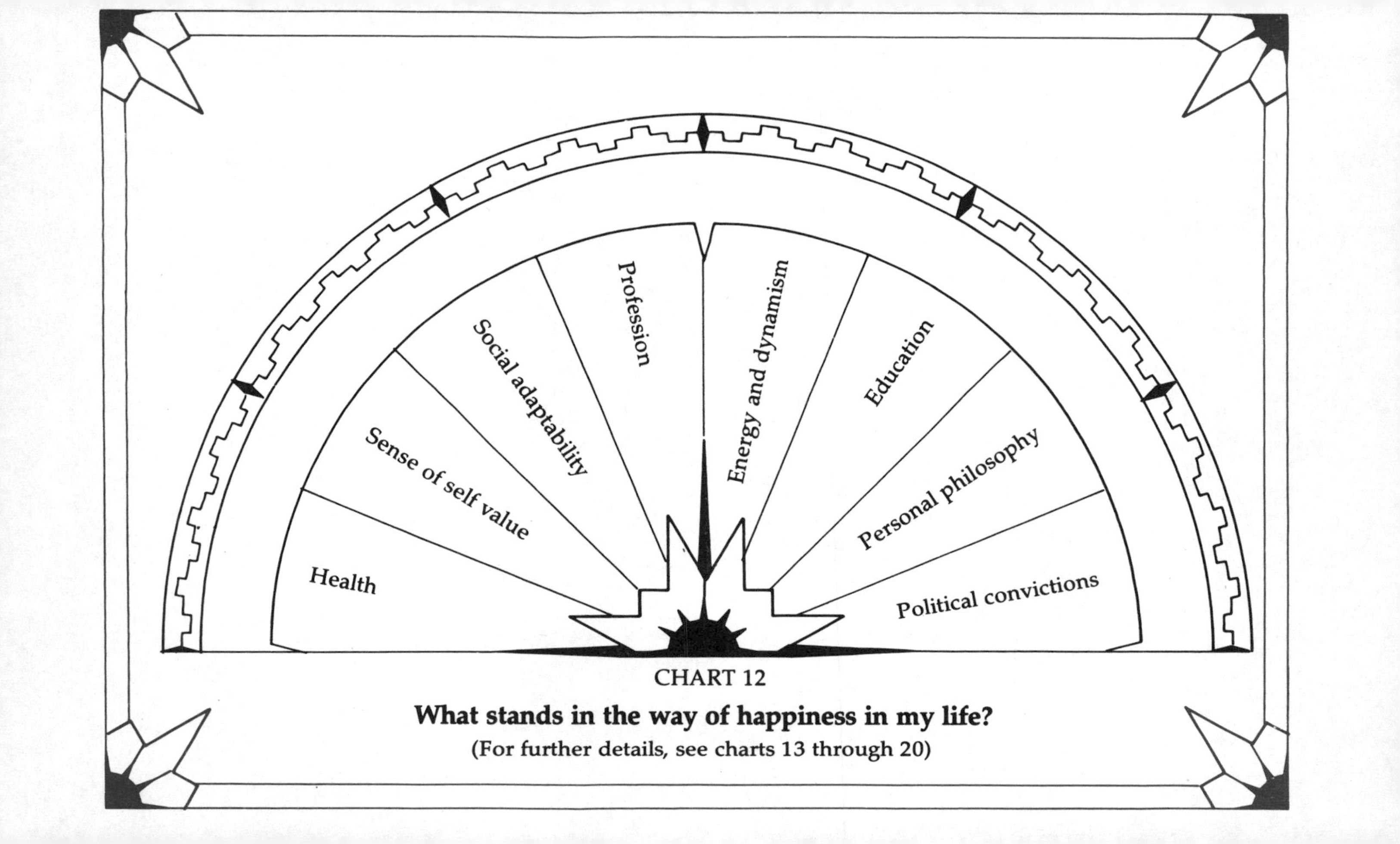

CHART 12

What stands in the way of happiness in my life?

(For further details, see charts 13 through 20)

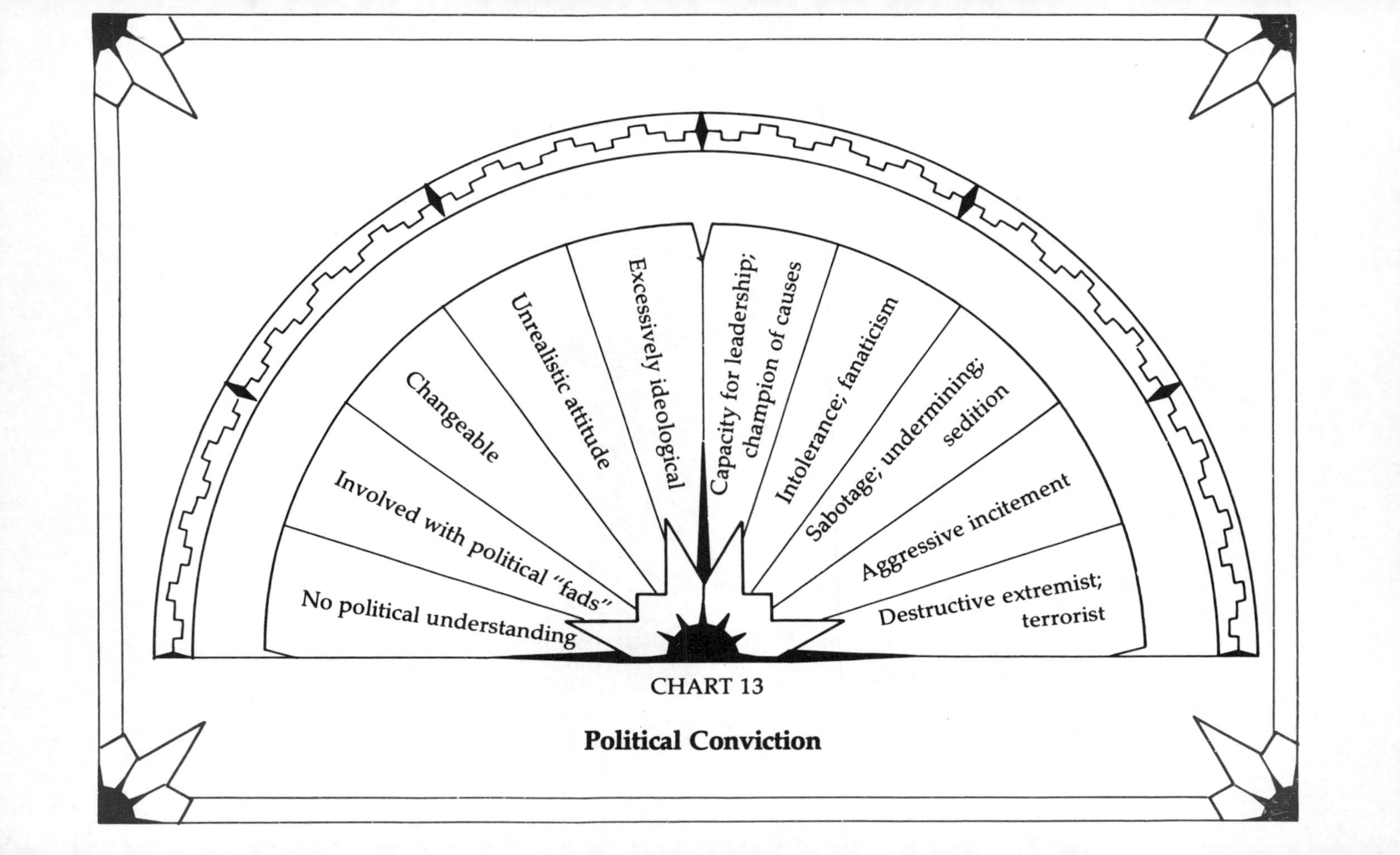

CHART 13

Political Conviction

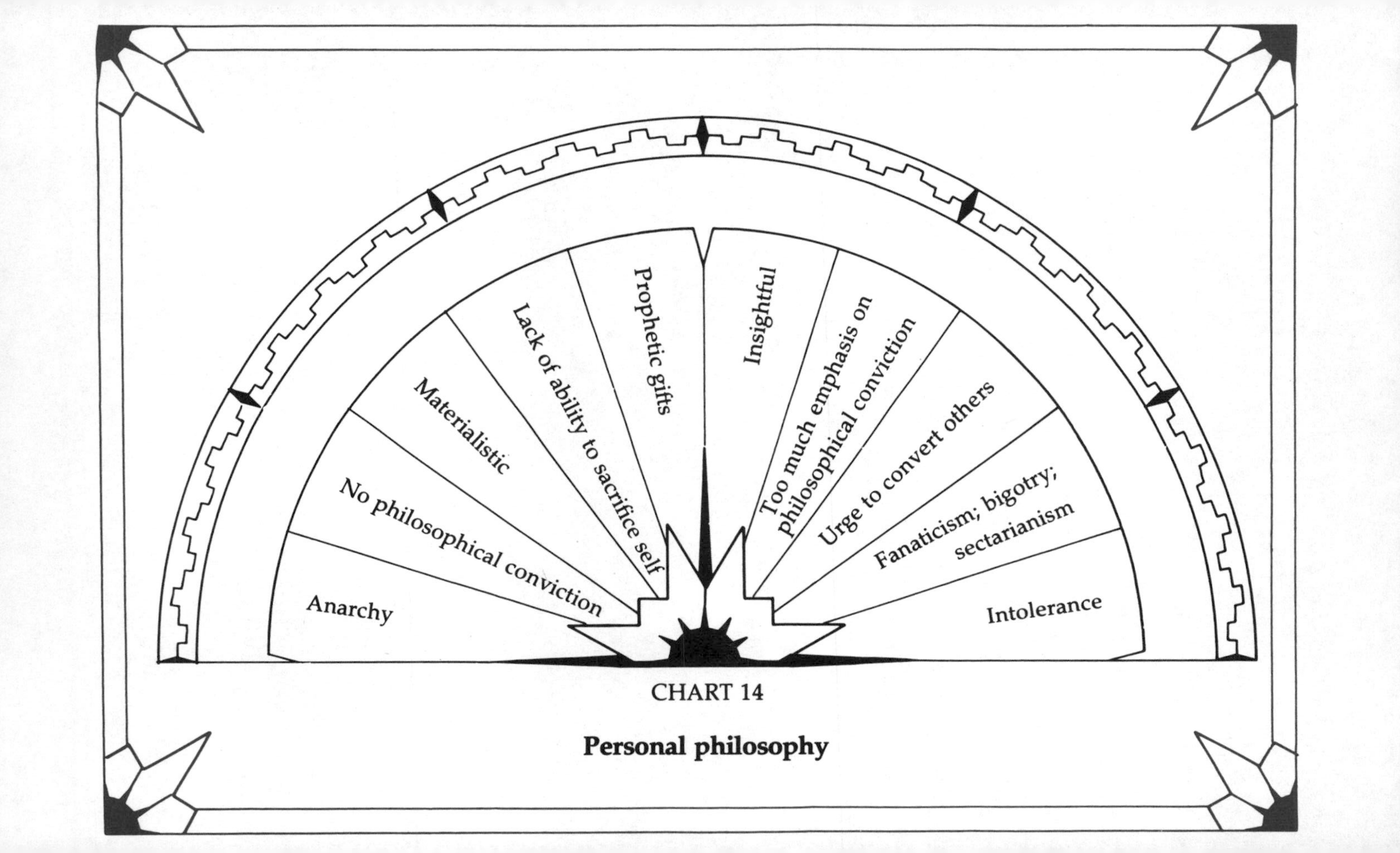

CHART 14

Personal philosophy

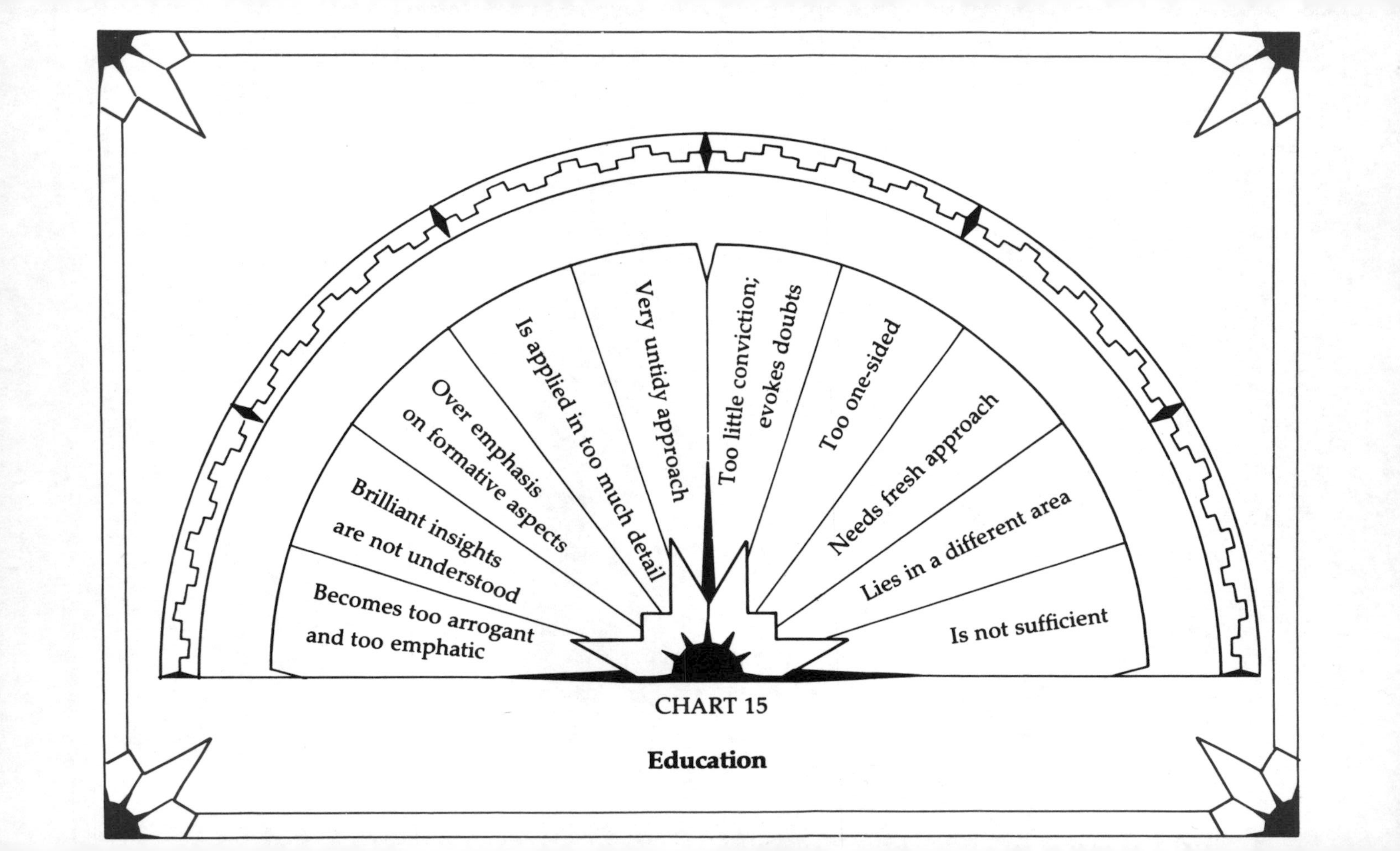

CHART 15

Education

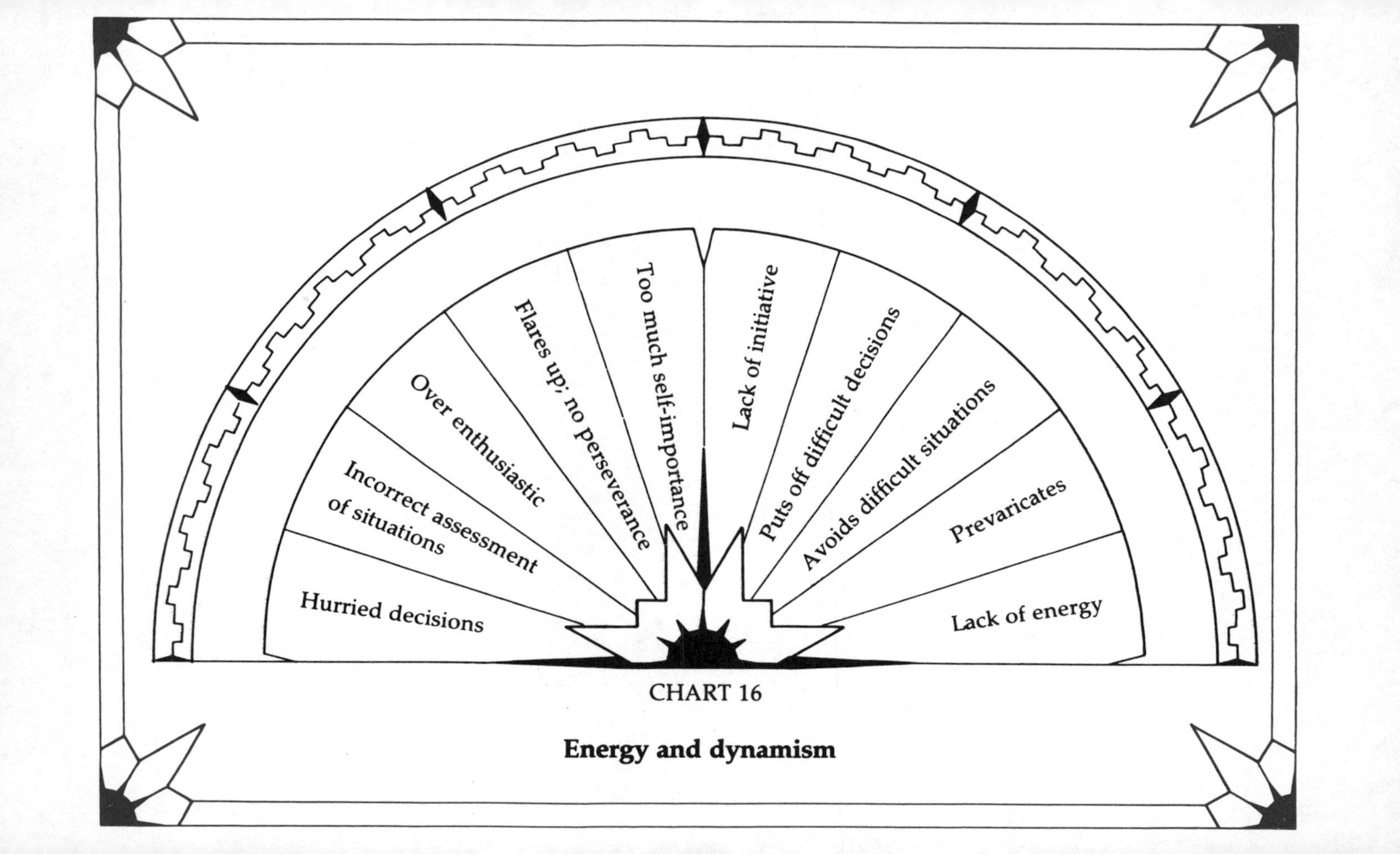

CHART 16

Energy and dynamism

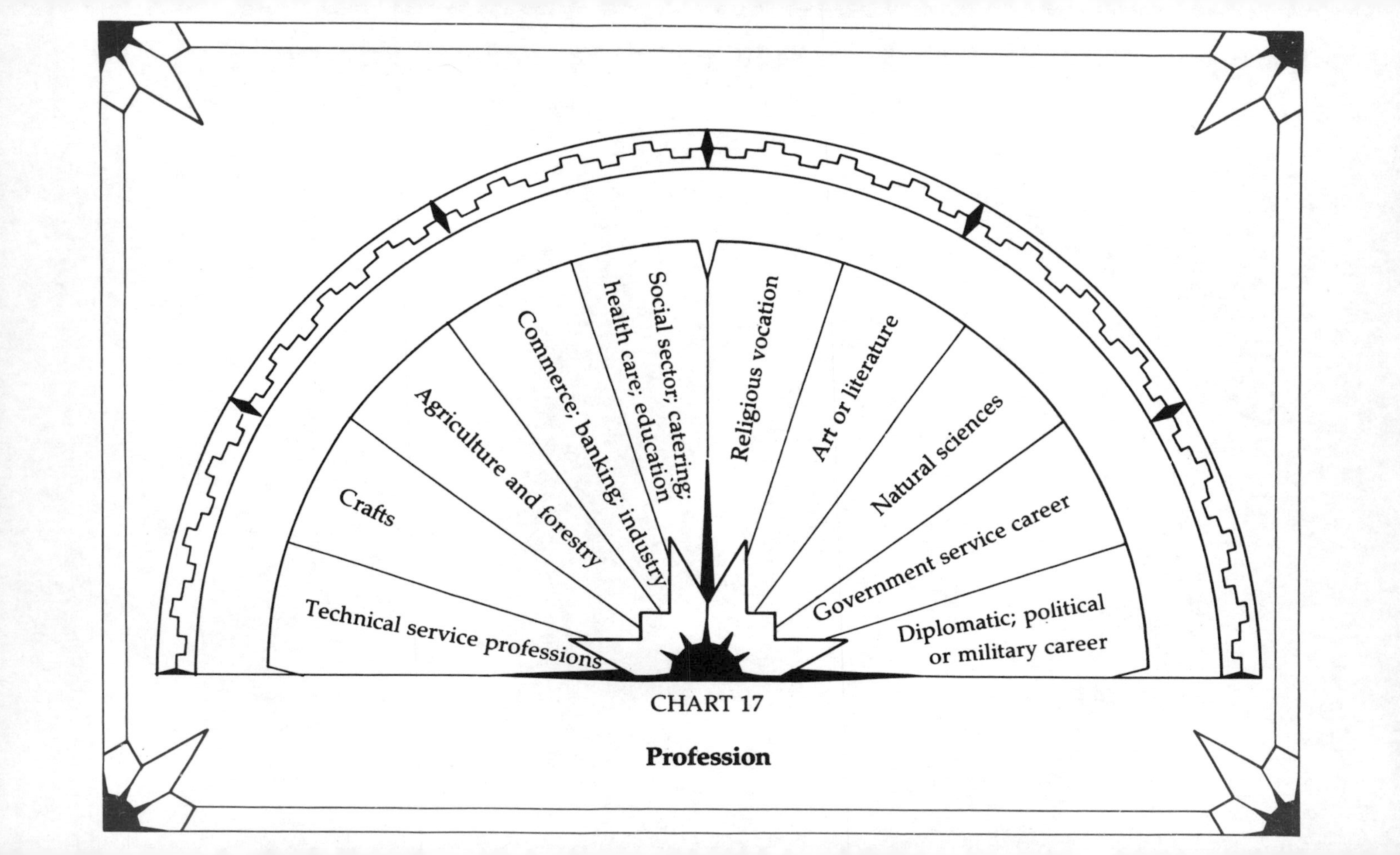

CHART 17

Profession

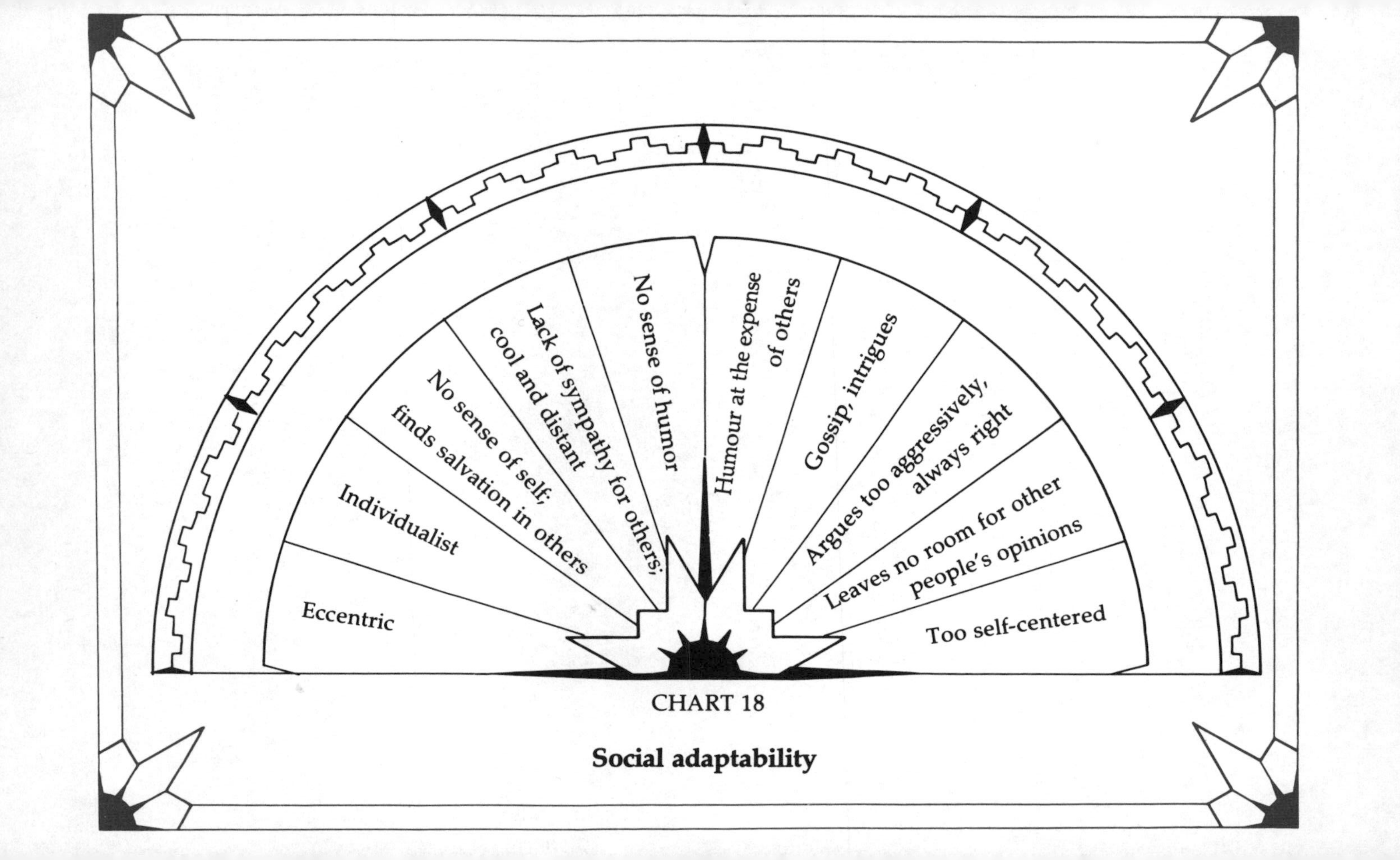

CHART 18

Social adaptability

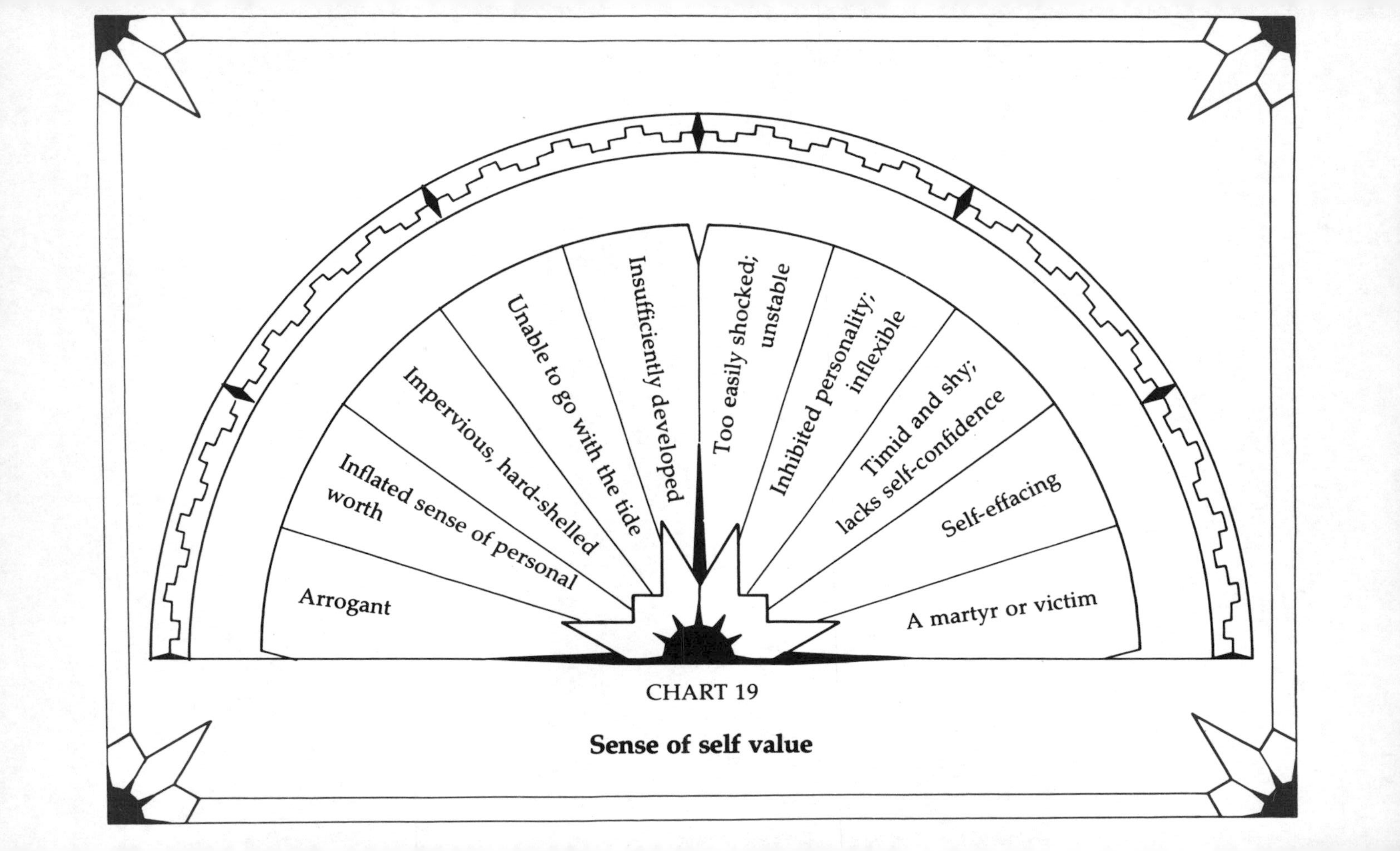

CHART 19

Sense of self value

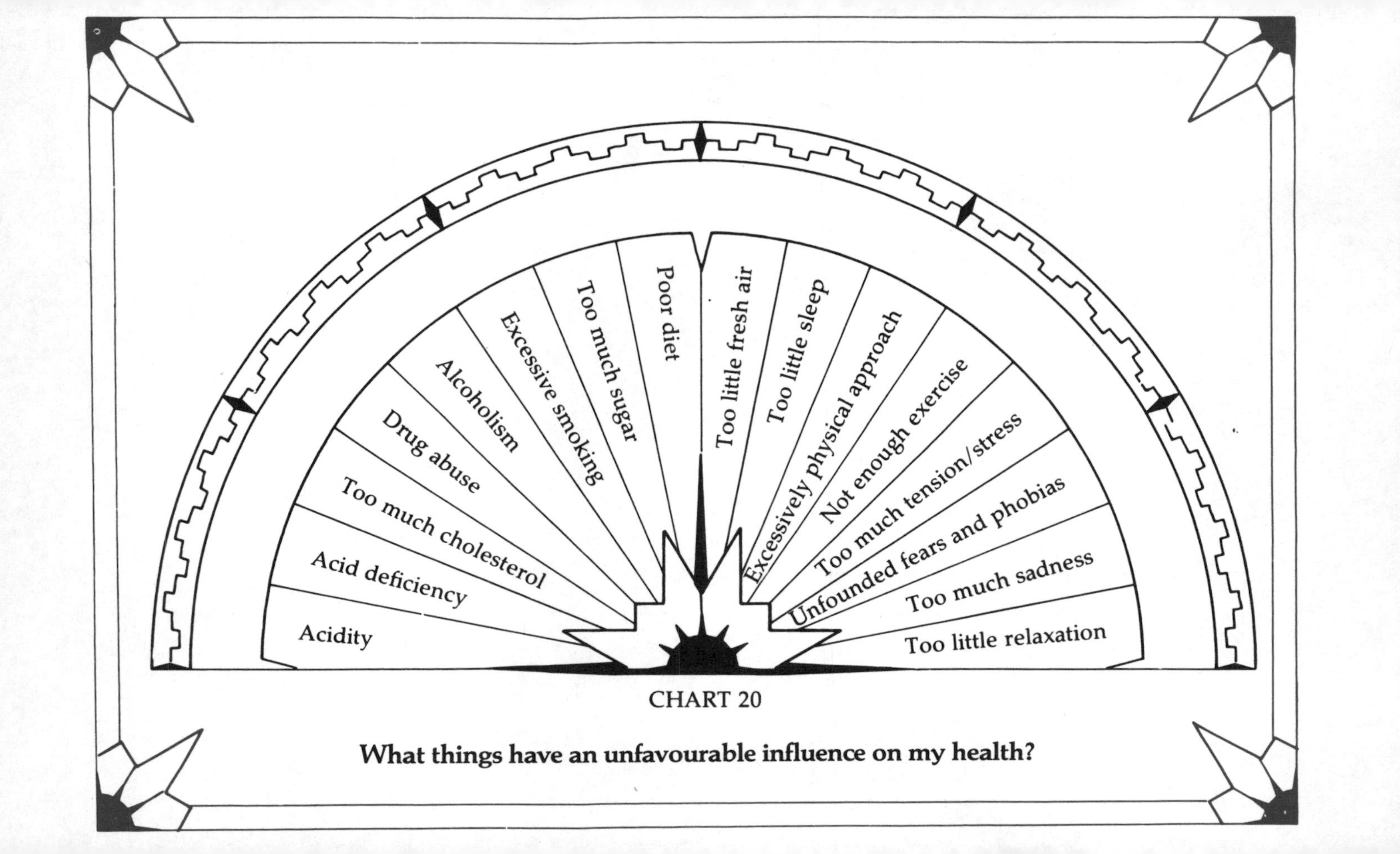

CHART 20

What things have an unfavourable influence on my health?

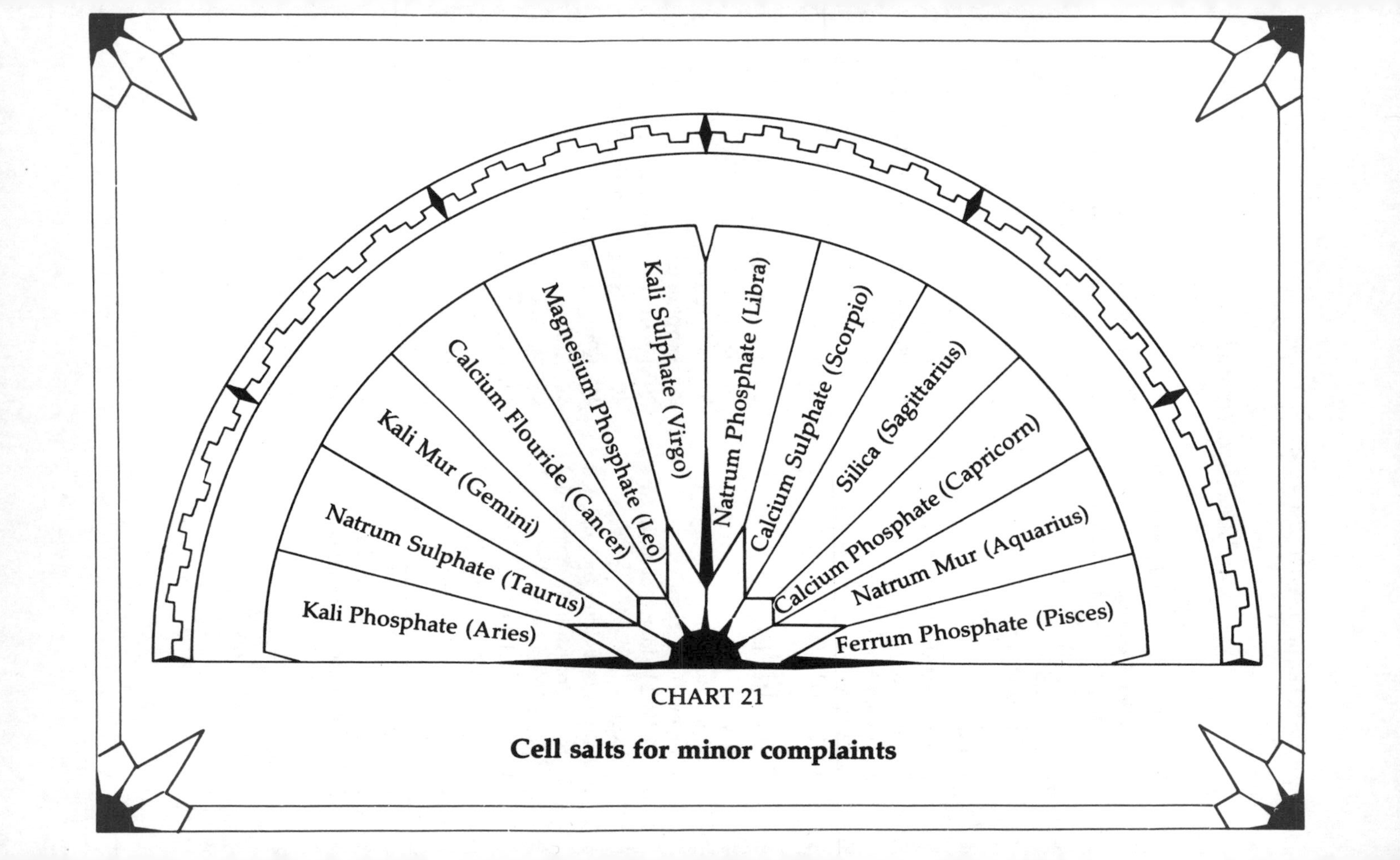

CHART 21

Cell salts for minor complaints

Location of Pain

1. Cortex, nervous centre. This regulates all the body's functions and the perception of everything in and around our bodies
2. Cerebral migraine
3. Eyes
4. Ears
5. Nasal cartilage
6. Jawbone
7. Tonsils
8. Larynx
9. Thyroid gland
10. Windpipe
11. Shoulder joint
12. Bronchi
13. Right lung
14. Liver, front right
15. Gall bladder, behind the liver
16. Pancreas, behind the stomach
17. Large intestine in horizontal position
18. Lower part of main artery
19. Lower part of aorta (hollow)
20. Large intestine upwards
21. Large intestine, downwards
22. Small intestine, coils
23. Appendix, front right
24. Worm-like appendage on appendix
25. Glands
26. Nails of the hands
27. Circulation; arteries from the heart through the body, and from parts of the body to the heart
28. Arteries/veins
29. Toenails
30. Hair
31. Pituitary in the middle of the cortex
32. Cerebellum
33. Spine (main nerve)
34. Aorta
35. Top of the lungs
36. Thymus
37. Left lung
38. Esophagus
39. Heart, between the lungs
40. Stomach, behind the liver
41. Elbow joint
42. Spleen
43. Kidneys
44. Pelvis
45. Urinary tracts
46. Hip joints
47. Bladder
48. Wrist joint
49. Genital organs
50. Knee joint
51. Ankle joint

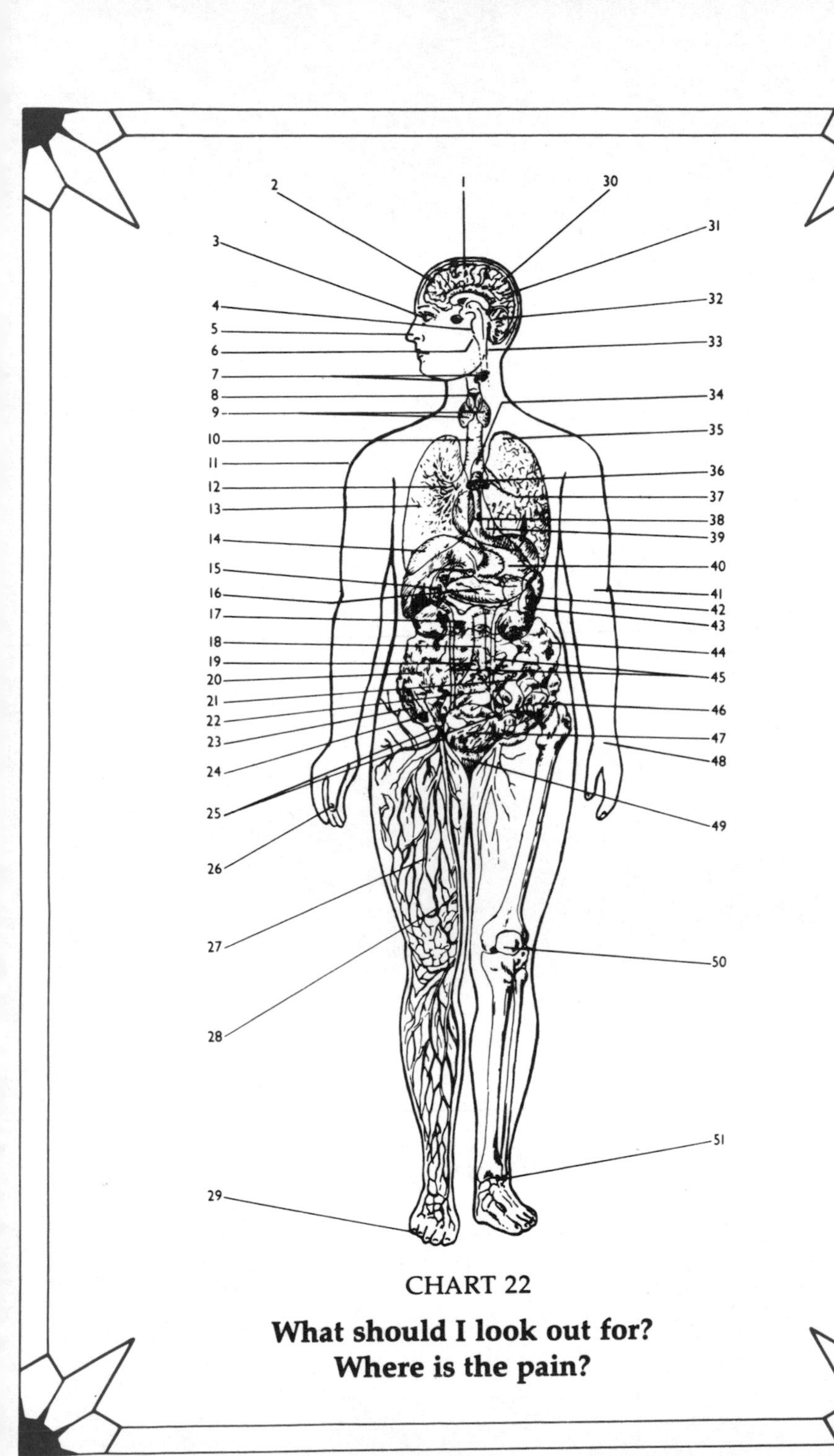

CHART 22

What should I look out for?
Where is the pain?

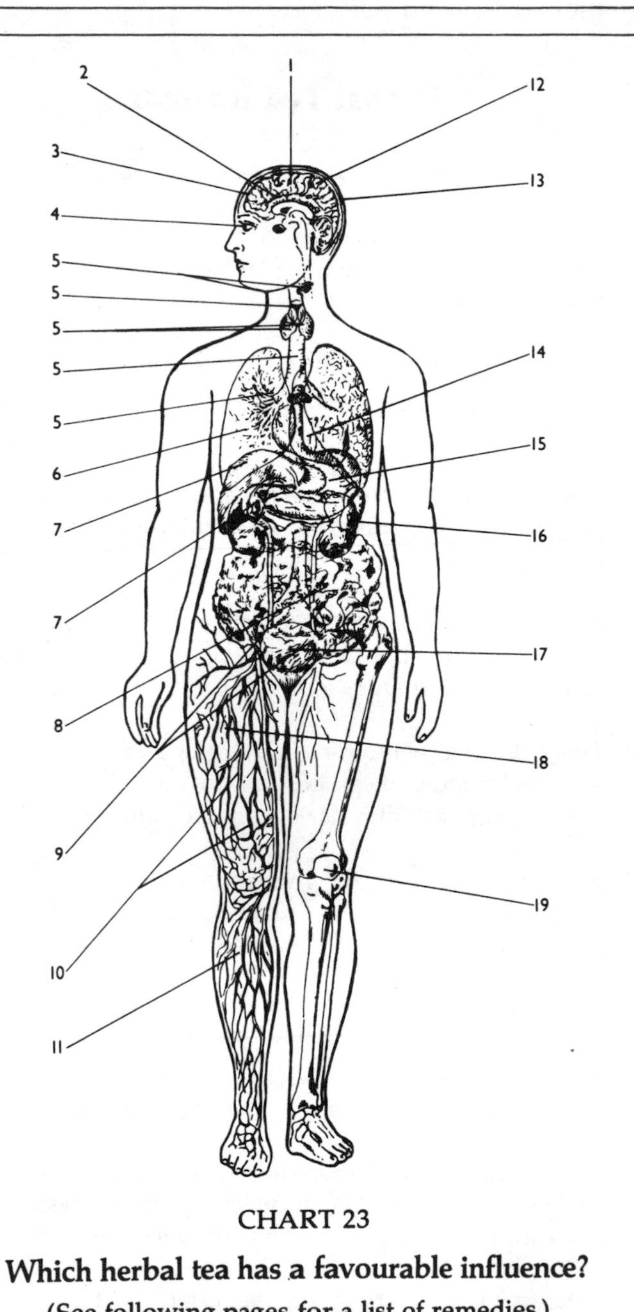

CHART 23

Which herbal tea has a favourable influence?

(See following pages for a list of remedies.)

Herbal Tea Remedies

1. *Cortex, nervous centre.* This perceives any disorder.

2. *Sleeplessness:* mint, passion flower, chamomile, garden sage, hops, motherwort, mullein, scullcap, catmint, lettuce, cowslip, linden blossom.

3. *Headaches:* mint, lavender, feverfew, thyme, rosemary, chamomile, garden sage, rhubarb, angelica, rue, marjoram, wood betony.

4. *Eyebath:* sassafrass, honey clover, angelica, rosemary, chamomile, hyssop, fennel, raspberry leaves, elderflowers, borage, golden seal, eyebright.

5. *Tonsils, larynx, colds, bronchi, sore throat:* garden sage, saffron, fennel, thyme, liquorice root, comfrey, golden seal, ginger, lungwort, horehound, vervain.

6. *Lungs:* lobelia, white pine bark, ginger, mullein, golden seal, hyssop/horehound, garlic, yarrow/elderflowers/peppermint, yarrow, slippery elm bark, vervain, comfrey, coltsfoot.

7. *Liver and gall bladder:* gentian root, bitterroot, bloodroot, bittersweet root, celery root and seed, self-heal, uva ursi, dandelion root and leaves, chamomile, angelica, golden seal, agrimony.

8. *Intestines:* Liquorice root, linseed, St. John's wort, slippery elm bark, golden seal, agrimony, blackberry leaves, marshmallow root, chamomile, scullcap, valerian, senna leaf and pod.

9. *Glands:* dandelion, lemon/honey, bayberry bark, bittersweet, yellow dock, poke root, mullein, agrimony/raspberry leaves, vervain, golden seal, scullcap, slippery elm bark, echinacea.

10. *Circulation (hardening of the arteries):* hawthorn, nettles, limeblossom, buckwheat, gentian, golden seal, cayenne.

11. *Cramp:* see *purification of the blood.*

12. *Nervous disorders:* garden sage, hops, wood betony, thyme, rue, lemon balm, chamomile, scullcap, catmint, peppermint, woodruff, velarian root, rosemary, lavender, mugwort.

13. *Hair loss:* Indian hemp, nettle, rosemary, garden sage, burdock, chamomile, spanish fly.

14. *Heart:* wood betony, bloodroot, mistletoe, holy thistle, tansy, peppermint, sorrel, coriander, hawthorn, lily of the valley, meadowsweet, slippery elm bark, angelica, borage, cayenne, golden seal, motherwort, gentian root.

15. *Stomach:* peppermint, centaury, yarrow, anise, fennel, gentian root, rosemary, slippery elm bark, wild yam, chamomile, marshmallow root, cayenne, tansy, bayberry bark, comfrey, thyme, bay leaves, marjoram, balomy, balm, ginger, dandelion root, garden sage, angelica.

16. *Spleen:* see *purification of the blood.*

17. *Bladder and urinary tract:* horsetail, gentian root, parsley, sweet balm, chicory, comfrey, celery, mugwort, squaw vine, hops, marjoram, tansy, clivers, couch grass, burdock root, St. John's wort, carrot tops, uva ursi, gravel root, marshmallow root, buchu leaves, corn silk, dandelion root, meadowsweet.

18. *For purification of the blood:* sorrel, dandelion root, blackthorn, prickly ash, spikenard, sassafrass, nettle, white clover, sarsaparilla, gentian root, bittersweet, parsley, chickweed, echinacea, golden seal, burdock root, smart weed, red clover, clivers, hyssop, St. John's wort, yellow dock, borage, tormentil.

19. *Rheumatism:* buchu leaves, catmint, wintergreen, yellow dock, burdock, mugwort, valerian, meadowsweet, corn silk, celery seed and root, red clover, garden sage, chamomile, wood betony.

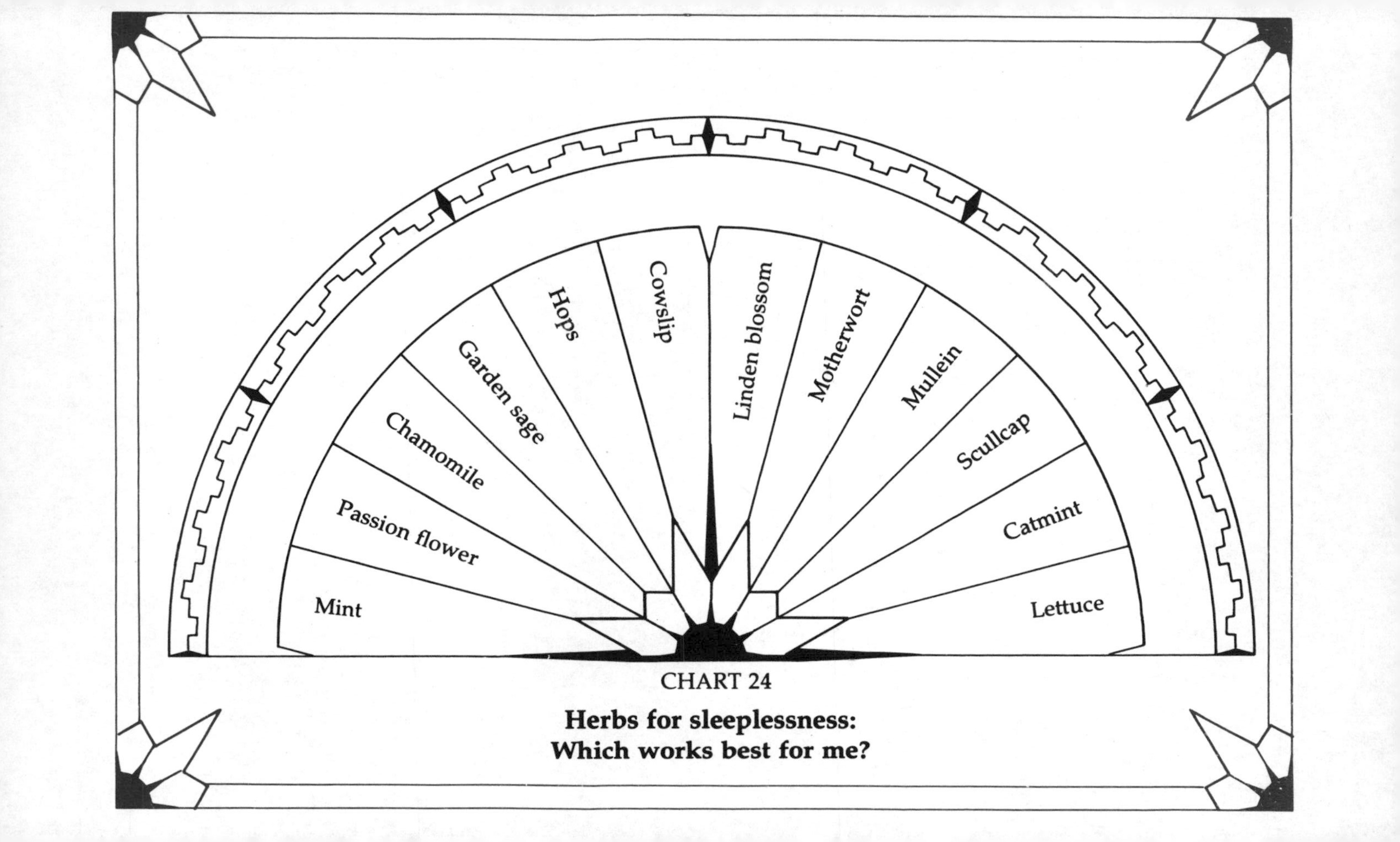

CHART 24

**Herbs for sleeplessness:
Which works best for me?**

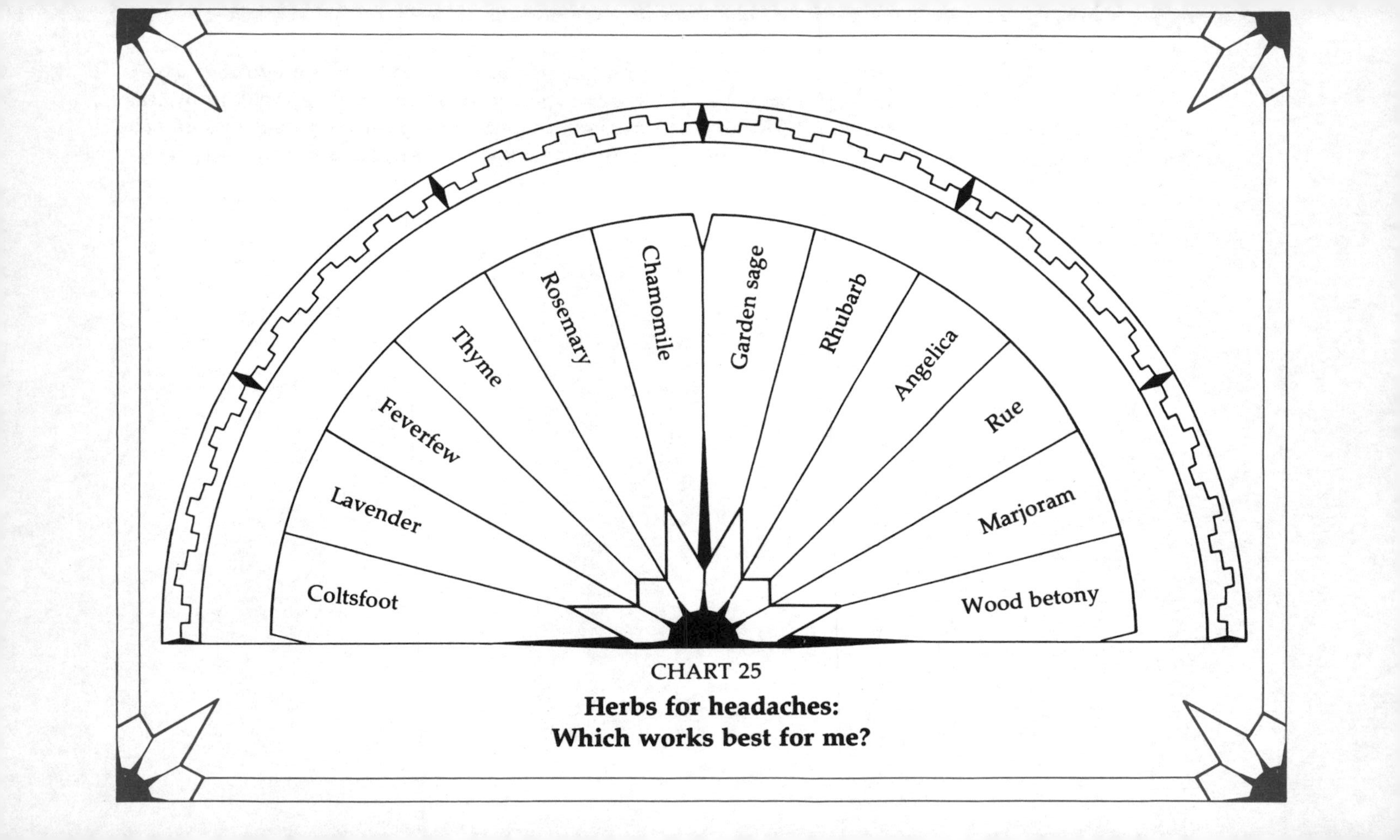

CHART 25

**Herbs for headaches:
Which works best for me?**

Note: These herbal remedies are used externally as compresses placed on tired and aching eyes. Best way to prepare them: Pour a cup of boiled spring water over one teaspoon of the herb and let steep until it cools. Then strain the liquid through a coffee filter or muslin. Soak pads of cotton in the liquid, and place over the eyes.

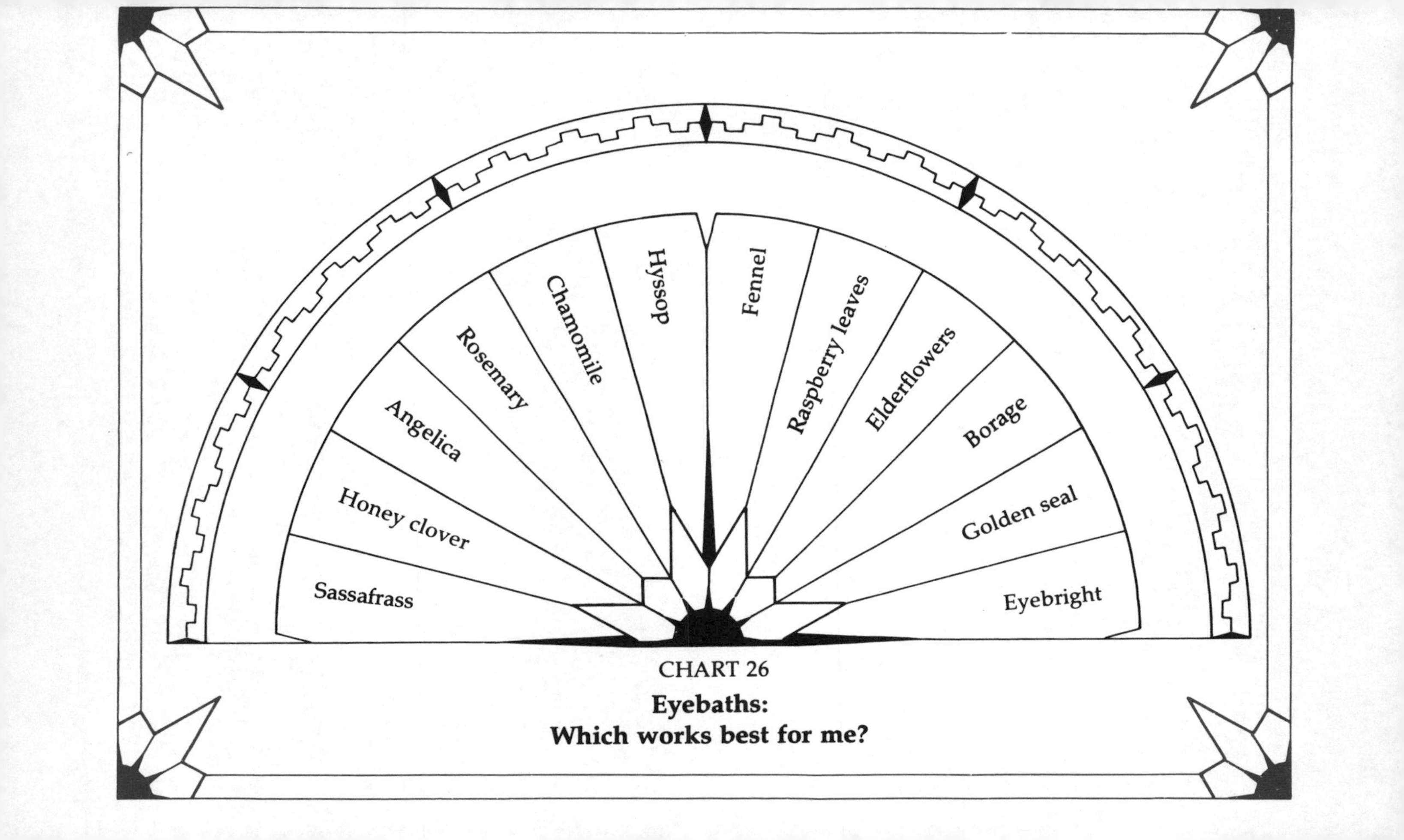

CHART 26

Eyebaths:
Which works best for me?

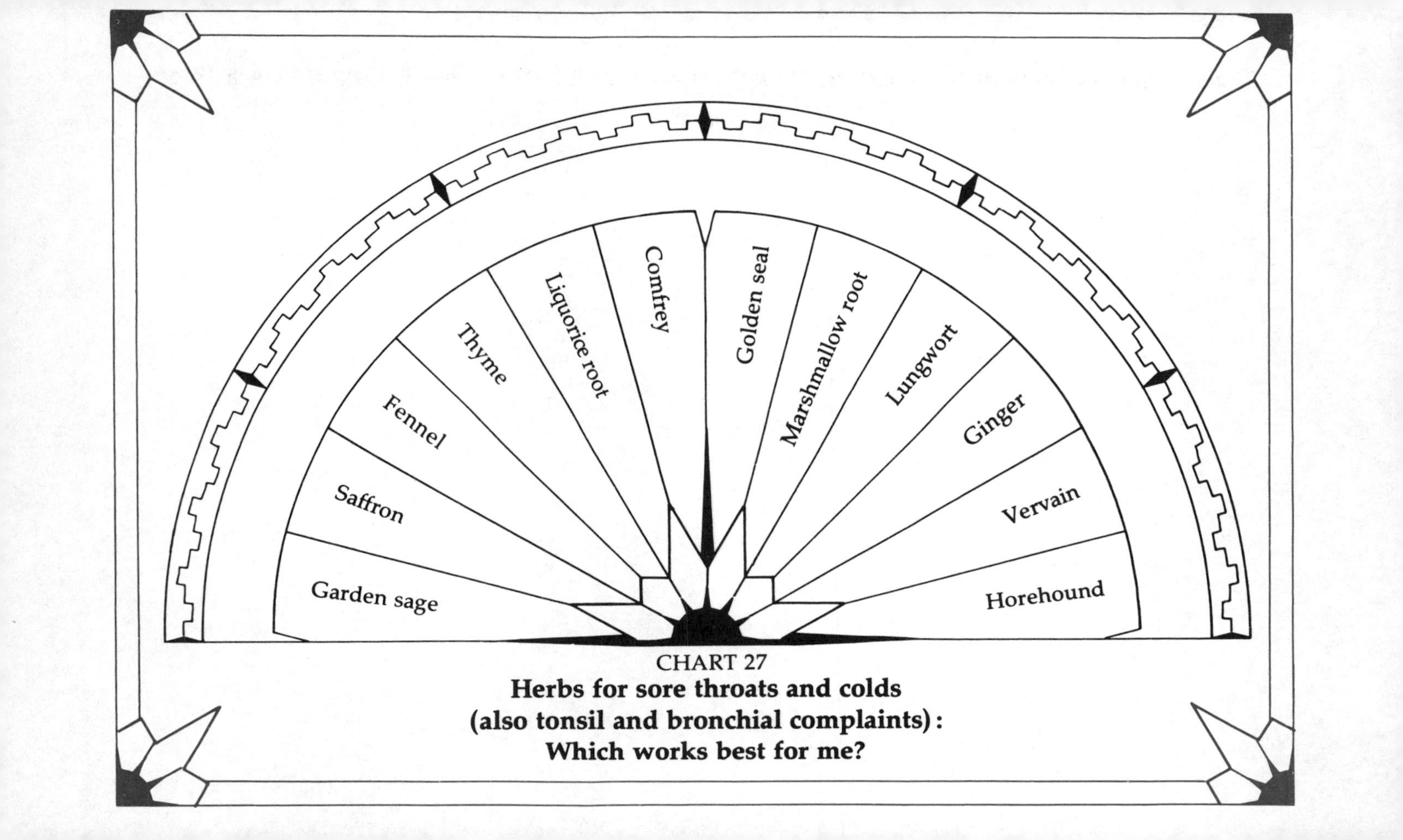

CHART 27
Herbs for sore throats and colds (also tonsil and bronchial complaints): Which works best for me?

Note: Remedies that are listed in combination are infusions made with equal parts of the herbs listed.

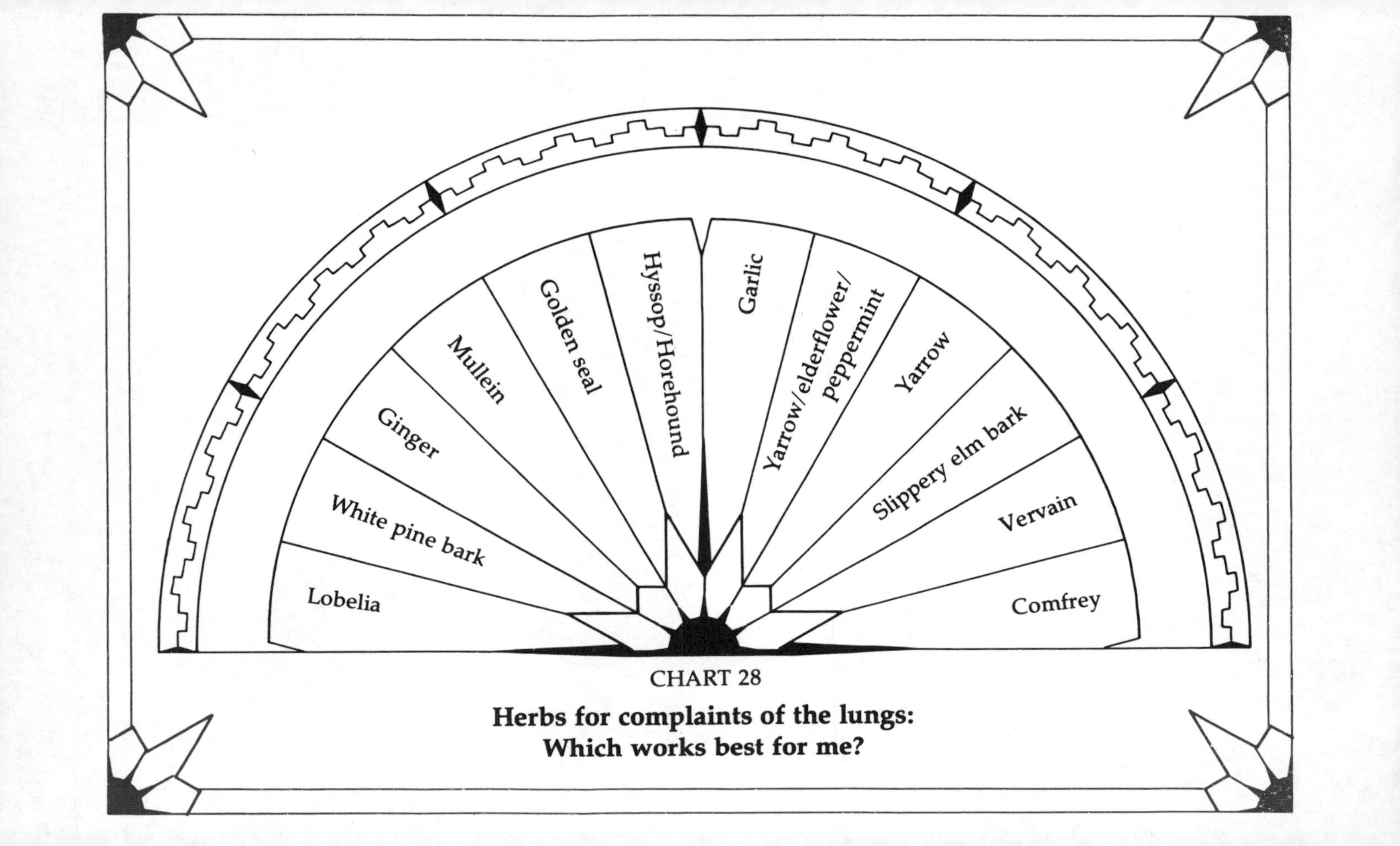

CHART 28

Herbs for complaints of the lungs:
Which works best for me?

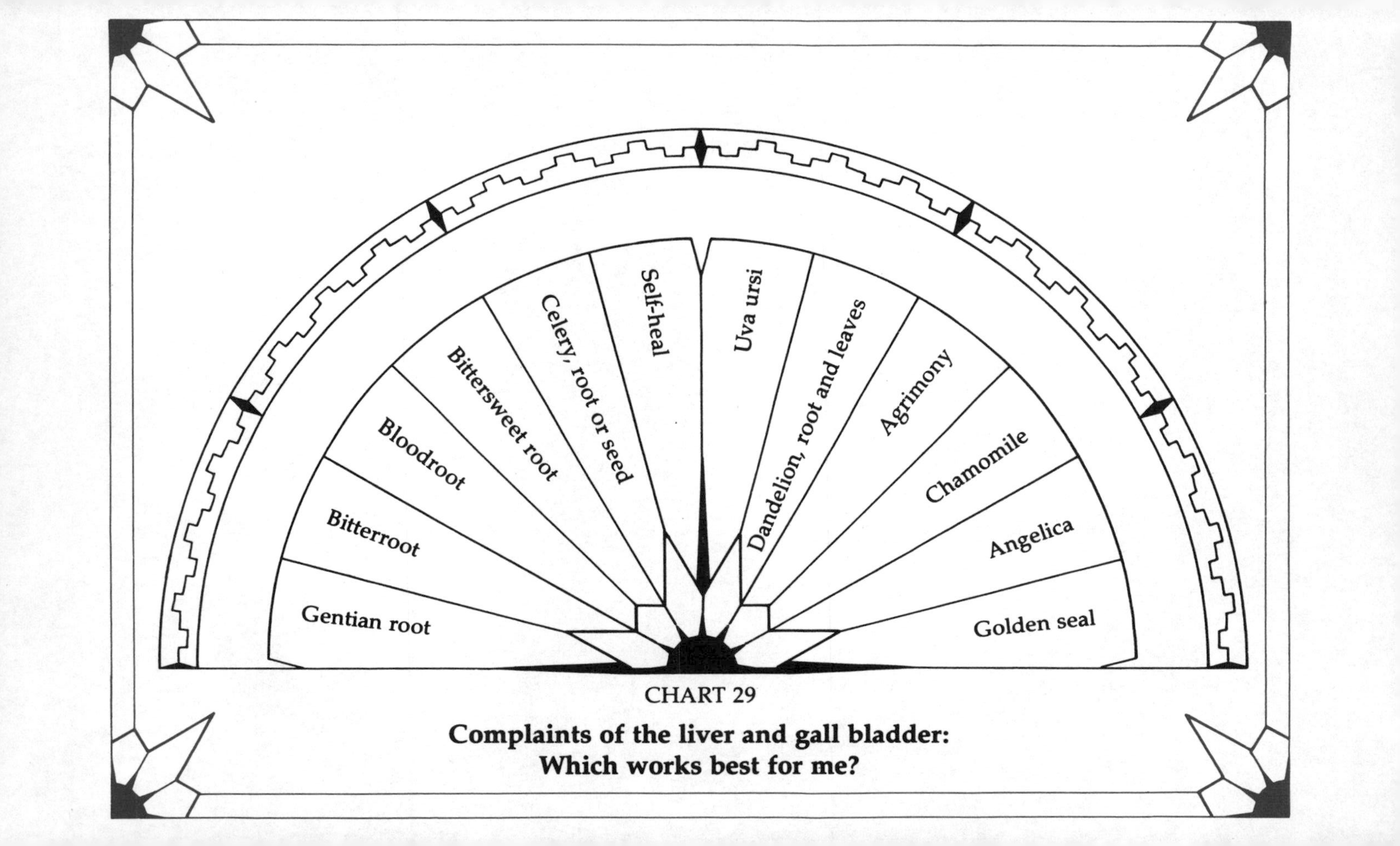

CHART 29

Complaints of the liver and gall bladder: Which works best for me?

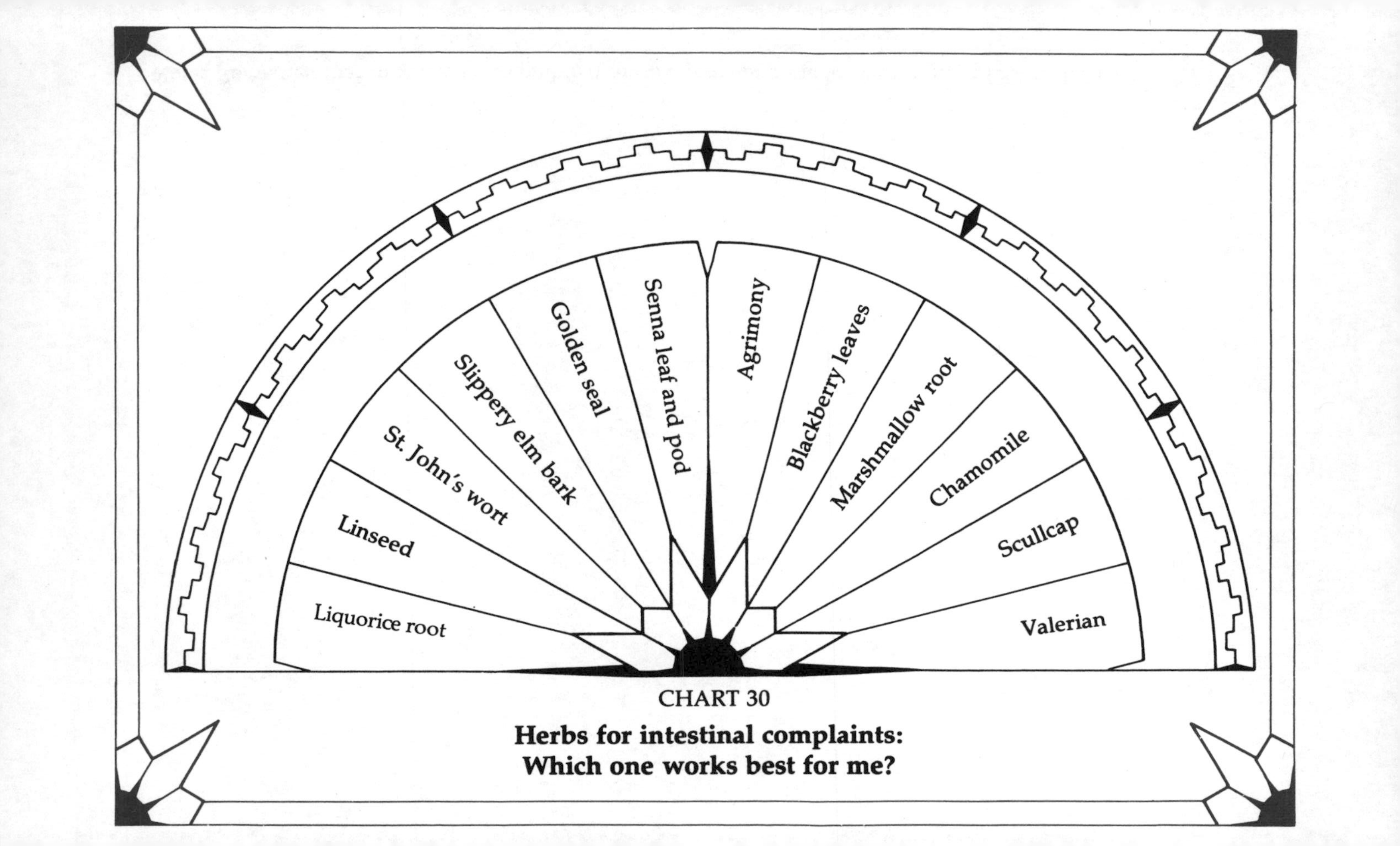

CHART 30

Herbs for intestinal complaints: Which one works best for me?

Note: Remedies that are listed in combination are infusions made with equal parts of herbs listed.

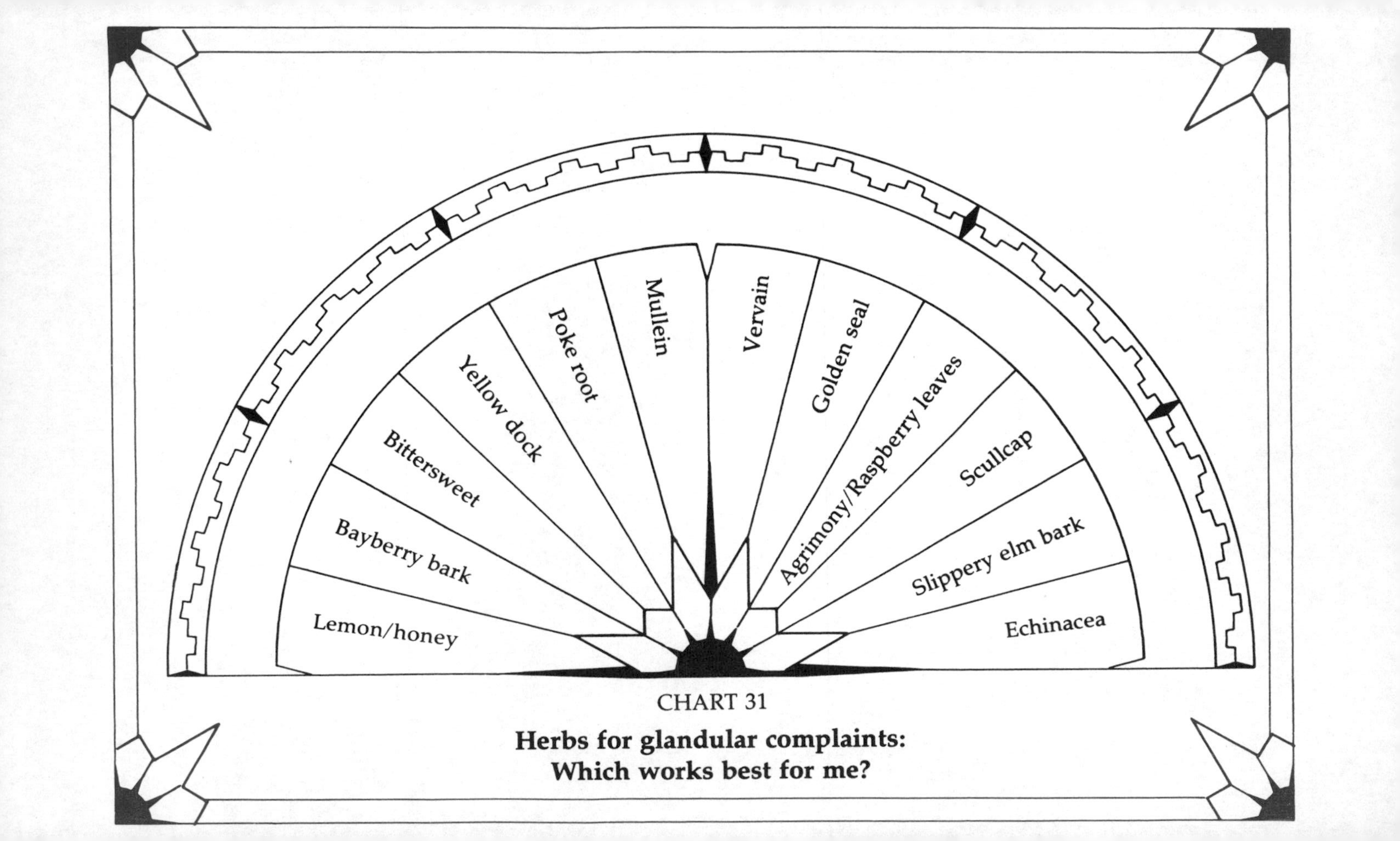

CHART 31

Herbs for glandular complaints:
Which works best for me?

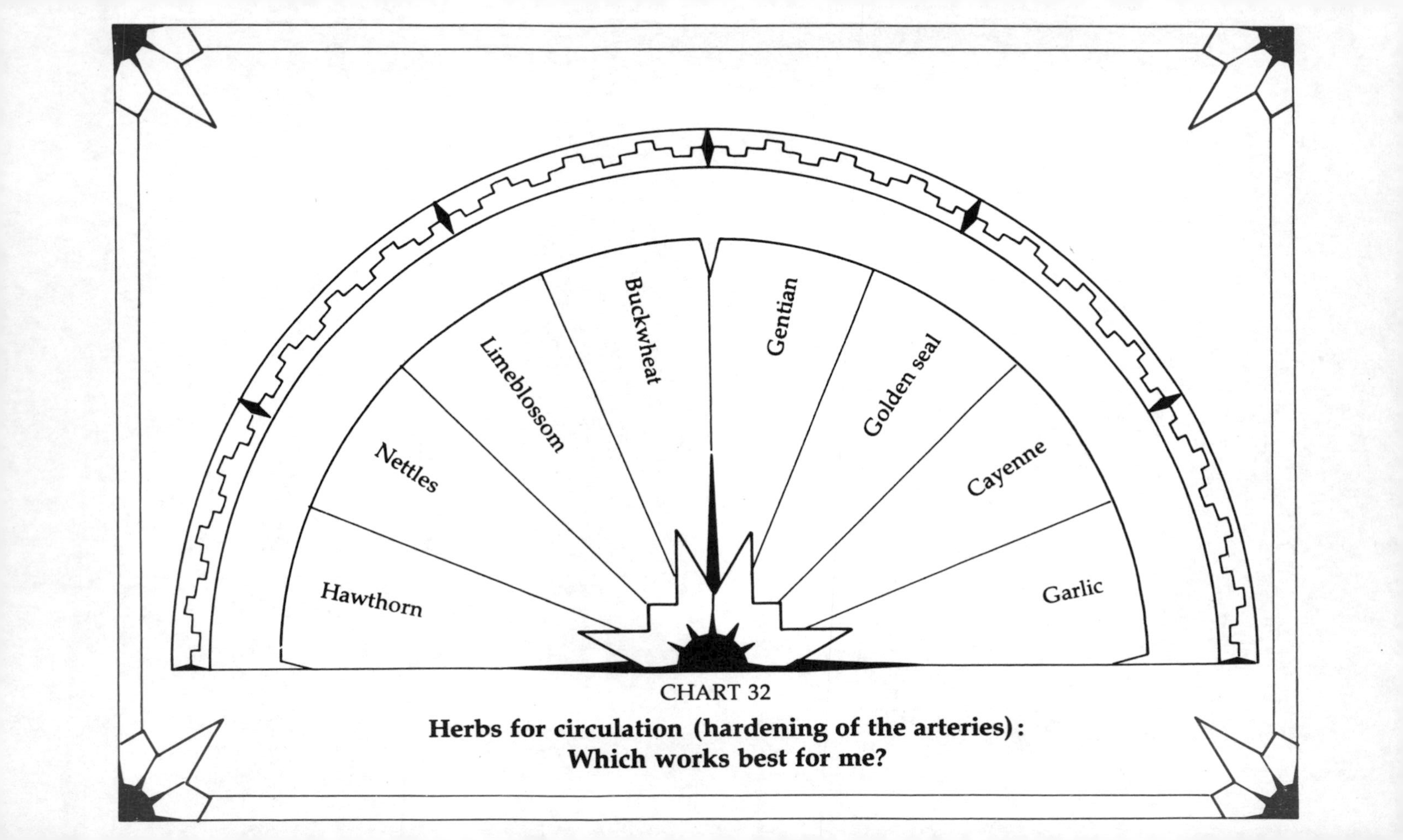

CHART 32

Herbs for circulation (hardening of the arteries): Which works best for me?

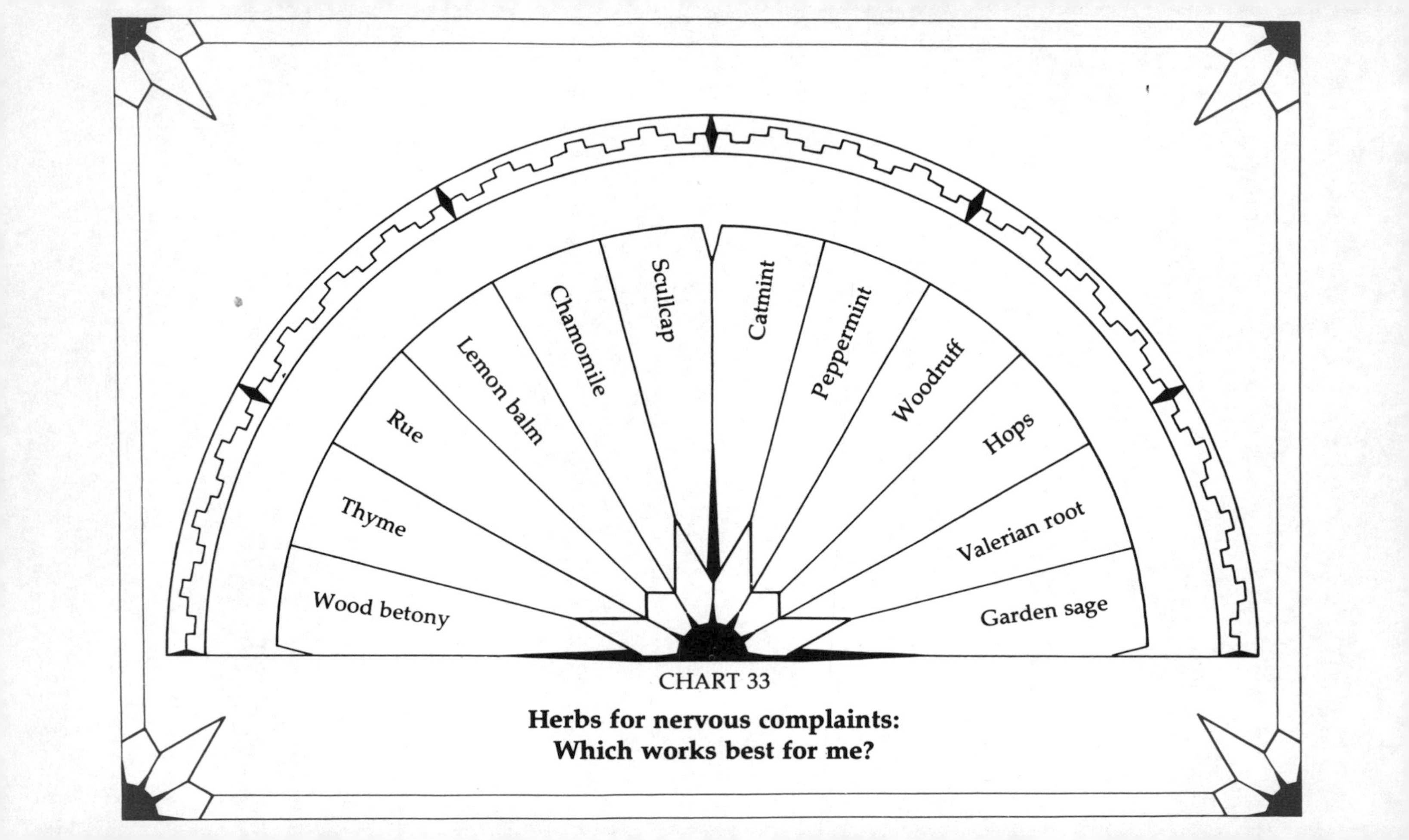

CHART 33

**Herbs for nervous complaints:
Which works best for me?**

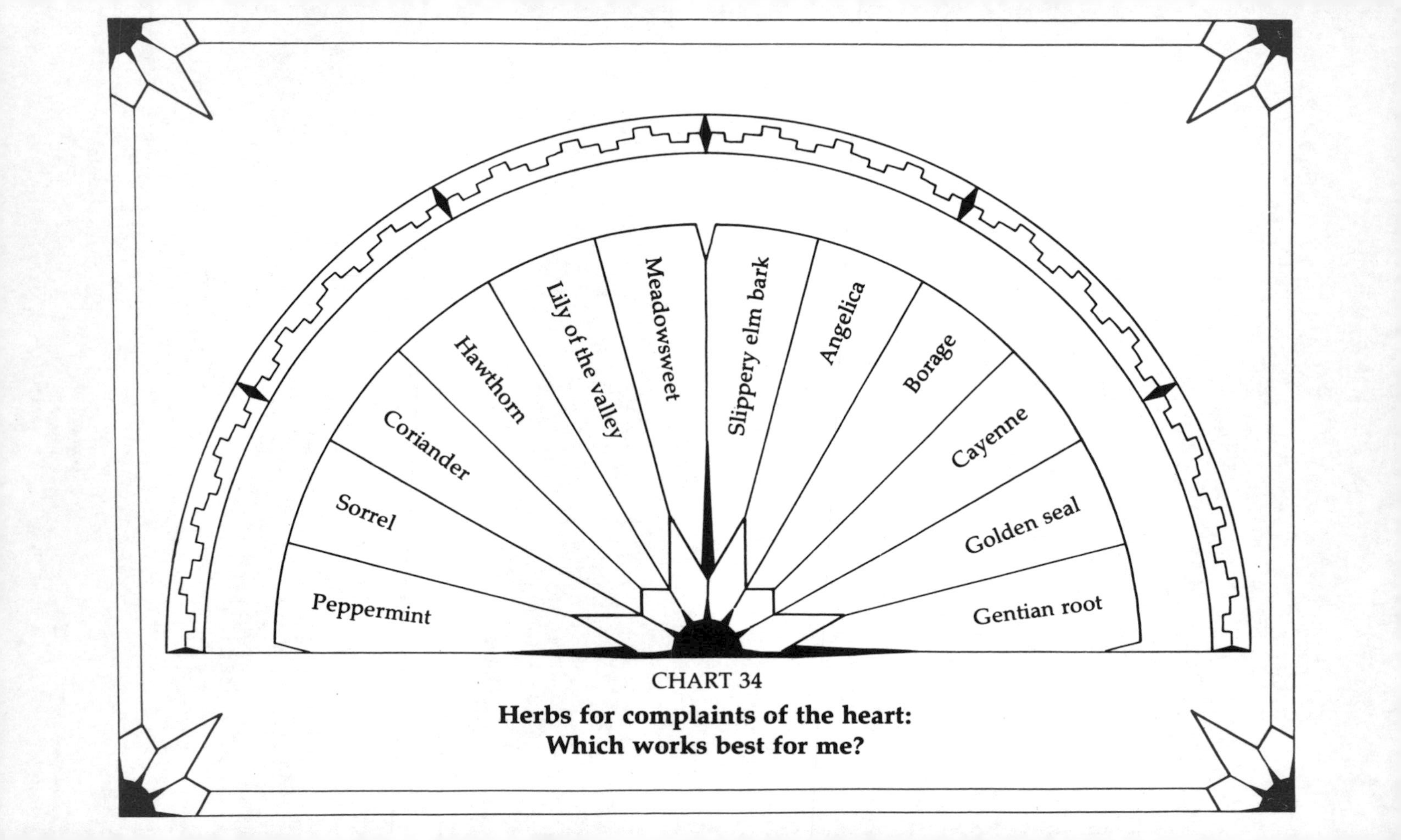

CHART 34

Herbs for complaints of the heart: Which works best for me?

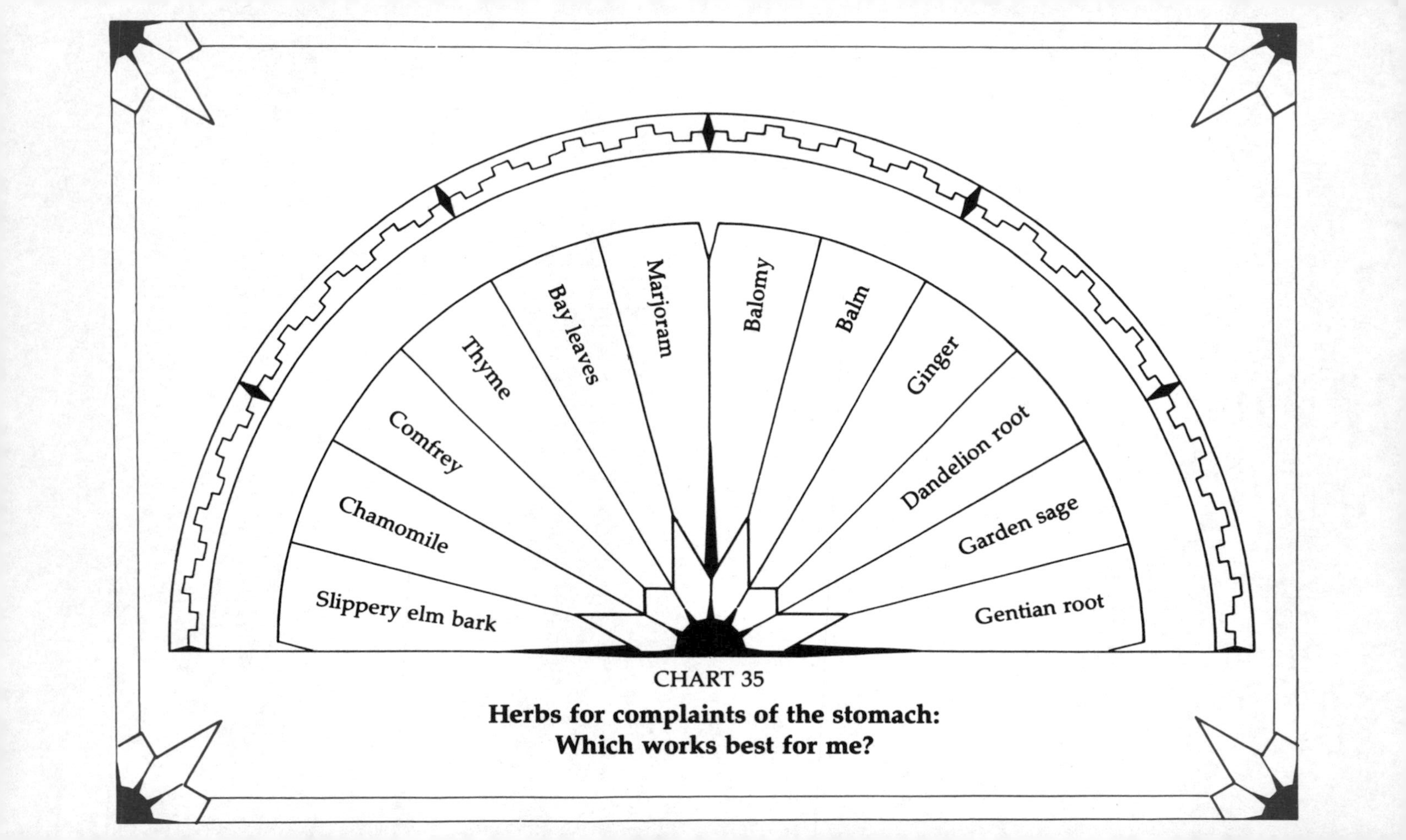

CHART 35

Herbs for complaints of the stomach:
Which works best for me?

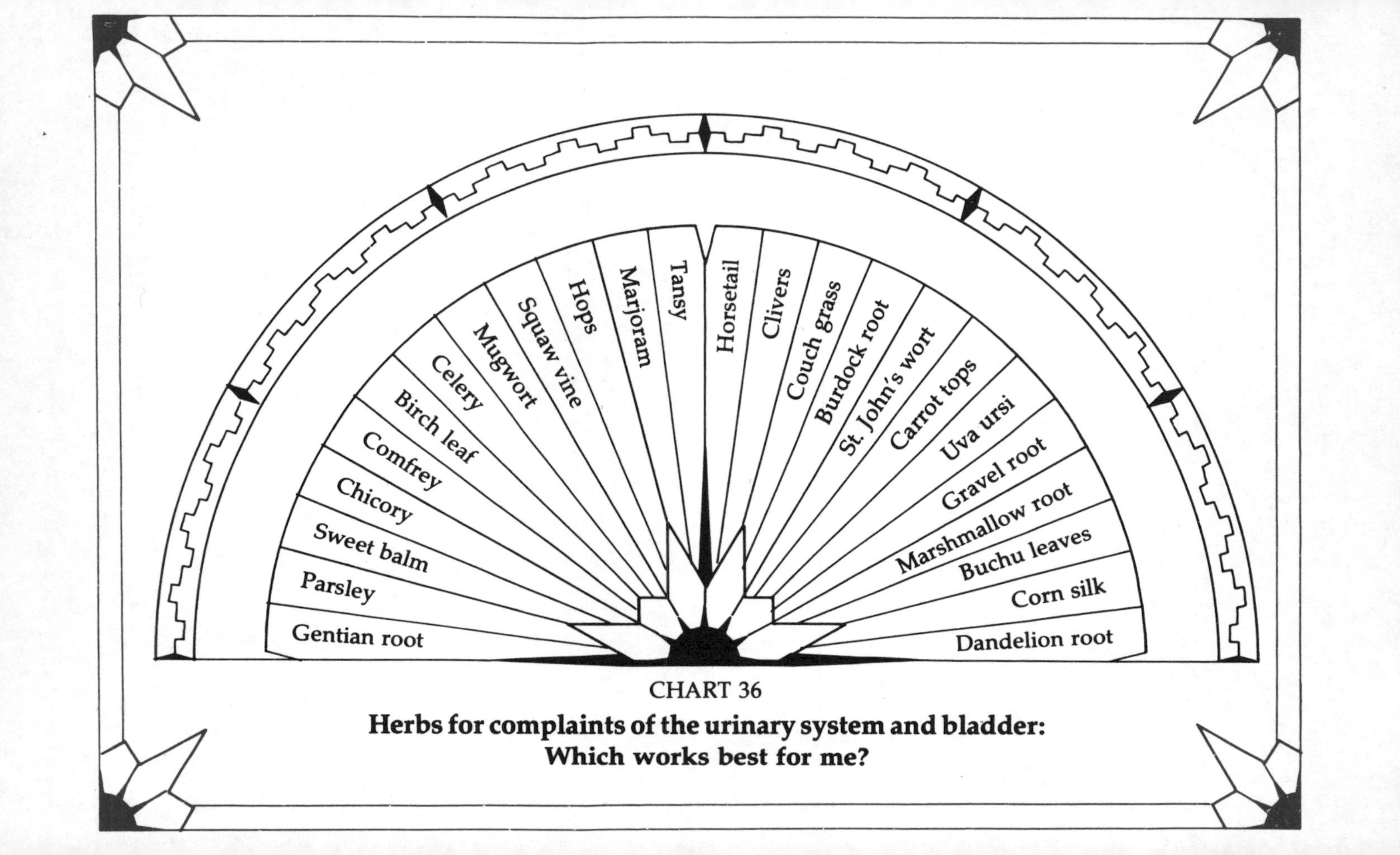

CHART 36

Herbs for complaints of the urinary system and bladder: Which works best for me?

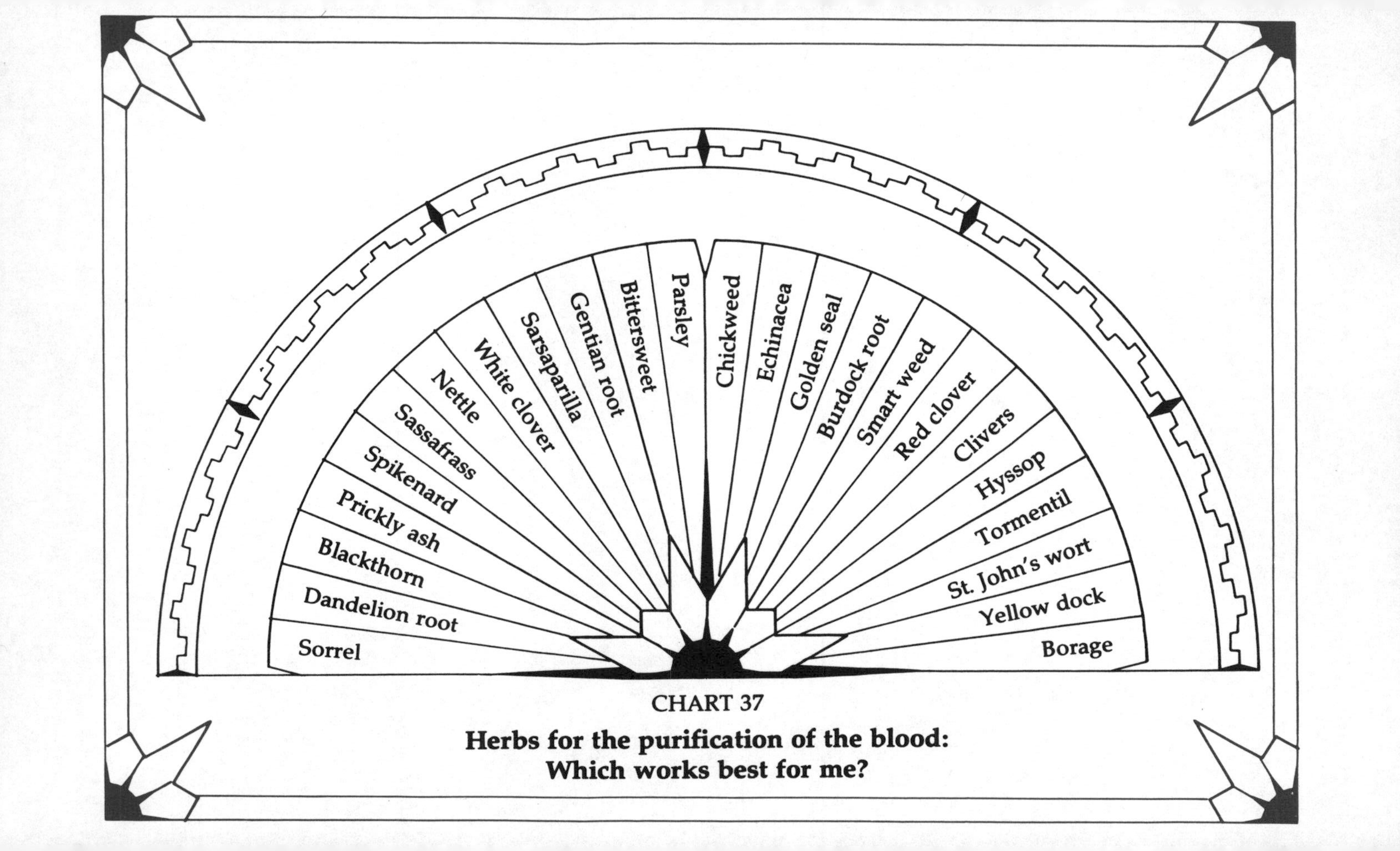

CHART 37

Herbs for the purification of the blood:
Which works best for me?

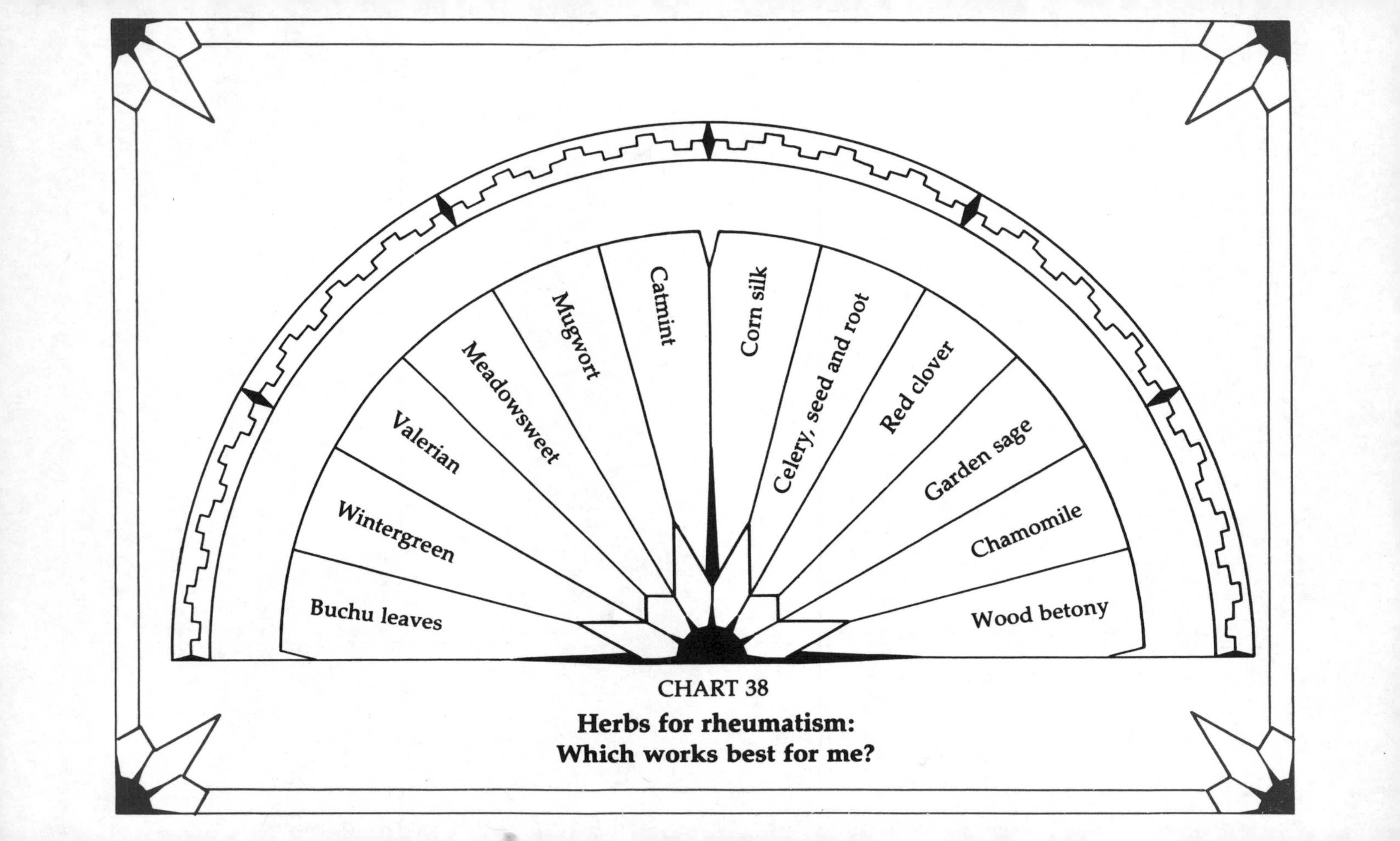

CHART 38

Herbs for rheumatism:
Which works best for me?